LABYRINTHS
OF REASON

Books by William Poundstone

BIG SECRETS
THE RECURSIVE UNIVERSE
BIGGER SECRETS
LABYRINTHS OF REASON

WILLIAM POUNDSTONE

LABYRINTHS

OF REASON

PARADOX, PUZZLES, AND THE FRAILTY OF KNOWLEDGE

ANCHOR BOOKS
DOUBLEDAY
NEW YORK   LONDON   TORONTO   SYDNEY   AUCKLAND

AN ANCHOR BOOK

PUBLISHED BY DOUBLEDAY

a division of Bantam Doubleday Dell Publishing Group, Inc.
666 Fifth Avenue, New York, New York 10103

ANCHOR BOOKS, DOUBLEDAY, and the portrayal of an anchor
are trademarks of Doubleday, a division of Bantam Doubleday
Dell Publishing Group, Inc.

This book was previously published in hardcover in 1988
by Anchor Books, an imprint of Doubleday.

Library of Congress Cataloging-in-Publication Data

Poundstone, William.
   Labyrinths of reason.

   Bibliography: p.
   Includes index.
   1. Science—Philosophy. 2. Paradox. I. Title.
Q175.P8816 1988   501   88-7066

To William Hilliard, Jr.

Contents

PART ONE

PART ONE

1

PARADOX

BLUE SKY, sunshine, déjà vu glazed with dread. Something horrible is going to happen about now. It is a perfect summer day in a meadow of tall grass. J.V. is following her brothers, lagging lazily behind. A shadow falls on the ground; something rustles the grass. J.V. turns—she cannot help it, it is what happens next—and sees a strange man. He has no face, like a minor character in a dream. The man holds something writhing and indistinct. He asks, "How would you like to get into this bag with the snakes?"

J.V.'s encounter is an unlikely milestone of twentieth-century thought. J.V., a fourteen-year-old girl, was not in a summer field but on an operating table in the Montreal Neurological Institute. Her physician, Wilder Penfield, was attempting an experimental operation to relieve her violent epileptic seizures. The operating team had removed the side of J.V.'s skull to expose the temporal lobe of the brain. In order to locate the site of the attacks, Penfield probed the brain with an electrode connected to an EEG machine. The surgery was a collaboration between physician and patient. J.V. had to remain conscious throughout and help locate the site of the seizures. When Penfield touched the probe to a certain spot on J.V.'s temporal lobe, she again found herself in the field of grass . . .

J.V.'s experience with the strange man had occurred seven years earlier, in Canada, in what we call the real world. She reported seeing herself as she was then, a seven-year-old girl. J.V. had been frightened but not physically harmed, and ran crying home to her mother. These few moments of terror were to haunt her over and over. The man with the bag of snakes entered her dreams, made them nightmares. The trauma became interwoven with her epileptic seizures. Like a madeleine, a fleeting recollection would trigger the whole memory, then an attack.

Under the EEG probe, J.V. not merely recalled but *relived* the encounter. All the richness of detail, all the lucid horror of the original experience, came back. Penfield's probe caused the brain to replay past experiences like a movie. With bits of lettered or numbered paper, Penfield kept track of the sites on the cerebral cortex associated with the recollection. Touching nearby points produced different sensations. When the probe touched one point, J.V. recalled people scolding her for doing something wrong. Other sites produced only a phantasmagoria of colored stars.

Brains in Vats

Penfield's classic brain experiments of the 1930s inspired a certain famous riddle, long since dubbed "brains in vats" by philosophy students. It goes like this: You think you're sitting there reading this book. Actually, you could be a disembodied brain in a laboratory somewhere, soaking in a vat of nutrients. Electrodes are attached to the brain, and a mad scientist is feeding it a stream of electrical impulses that exactly *simulates* the experience of reading this book!

Let's expand a little on the anecdote to see the full force of this. At some indistinct past time, while you were sleeping, your brain was removed from your body. Every nerve was severed by skilled surgeons and attached to a microscopic electrode. Each of these millions of electrodes is hooked to a machine that produces tiny electrical pulses just as the original nerves did.

When you turn the page, it *feels* like a page because the electrodes send your brain exactly the same nerve impulses that would have come from real fingers grasping a real page. But the page and the fingers are illusion. Bringing the book closer to your face makes it look bigger; holding it at arm's length makes it look smaller . . . 3-D perspective is simulated by judiciously adjusting the voltages of the electrodes attached to the stump of the optic nerve. If, right this instant, you can smell spaghetti cooking and hear dulcimer music in the background, that is part of the illusion too. You can pinch yourself and receive the expected sensation, but it will prove nothing. In fact, *there's no way you can prove that this isn't so.* How, then, can you justify your belief that the external world exists?

Dreams and Evil Geniuses

To anyone with a skeptical turn of mind, the brains-in-vats paradox is both appealing and infuriating. There is something fascinating about the demonstration that, just possibly, everything you know is wrong!

Despite the influence of Penfield and other brain researchers, doubts about the reality of the world are not a uniquely modern malaise. Brains-in-vats is simply a stronger version of older riddles asking "How do you know this isn't all a dream?" Best known of these is the Chinese tale of Chuang-tzu, dating from the fourth century B.C. Chuang-tzu was the man who dreamt he was a butterfly, then awoke to wonder if he was a butterfly dreaming he was a man.

Chuang-tzu's fable is unconvincing. It is true that we usually don't realize we're dreaming in our dreams. But a waking person always knows that he is not dreaming. Doesn't he?

Opinions differ. In his "First Meditation" (1641), French philosopher and mathematician René Descartes decided he could not be *absolutely* sure he wasn't dreaming. Most people would probably disagree with Descartes. You're not dreaming right now, and you know it because experiences in dreams are different from those in waking life.

Saying exactly *how* they're different is difficult. If waking life is absolutely, unmistakably different from a dream, there ought to be some surefire test you can perform to distinguish the two. For instance:

• There's the old gag about pinching yourself to see if you're dreaming. The rationale is apparently that you don't feel pain in dreams. But I *have* felt pain in dreams, and suspect that everyone must from time to time. Scratch that test.

• Since few dreams are in color, the red rose on your desk proves you're awake. Again, the dream sensation of color is not all that rare. Many people dream in color, and even if you never have, this could be the first time.

• Real life usually seems more detailed and coherent than dreams. If you can examine the wall before you and see every minute crack, that means you're awake. If you can add a column of figures, then check the result with a calculator, you're awake. These tests are more telling though still not foolproof. (Might not you dream about seeing tiny cracks in the wall after hearing that the cracks "prove" you are awake?)

• Some say that the very fact that you are wondering whether you are dreaming or awake proves you are awake. In waking life, you are aware of the dream state, but while dreaming you forget the distinction (and think you are awake). But if that were true, you could never have a dream in which you realize you are dreaming, and such dreams are fairly common with many people.

• I propose this test, based on what might be called "coherent novelty." Keep a book of limericks by your bed. Don't read the book; just use it thus. Whenever you want to know if you are dreaming, go into your bedroom and open the book at random (it may of course be a dream bedroom and a dream book). Read a limerick, making sure it is one you have never read or heard before. Most likely you cannot compose a bona fide limerick on a moment's notice. You can't do it when awake, and certainly not when asleep either. Nonetheless, anyone can *recognize* a limerick when he sees it. It has a precise rhyme and metrical scheme, and it is funny (or more likely *not* funny, but in a certain way). If the limerick meets all these tests, it must be part of the external world and not a figment of your dreaming mind.[1]

---

[1] Samuel Taylor Coleridge composed his masterwork, "Kubla Khan," in a dream. Coleridge fell asleep reading a history of the emperor and dreamed, with startling lucidity, a poem of 300 verses. Upon awakening, Coleridge scrambled to write down the poem before it eluded him. He wrote about 50 verses—the "Kubla Khan" we know—then was interrupted by a visitor. Afterward, he could remember but a few scattered lines of the remaining 250 verses. Coleridge, however, was a poet in waking life. I recommend the limerick test only to people who can't easily compose a limerick. Also, Coleridge's dream was perhaps atypical, for he had taken laudanum to get to sleep.

There was a young girl at Bryn Mawr
Who committed a dreadful faux pas;
   She loosened a stay
   Of her décolleté
Exposing her je-ne-sais-quoi.

My real point is that you don't need to use any of these tests to establish that you're awake; you just *know*. The suggestion that Chuang Tzu's, or anyone's, "real" life is literally a nighttime dream lacks credibility.

It may however be a "dream" of a different sort. The most famous discussion along these lines is in Descartes's *Meditations*. There Descartes wonders if the external world, including his body, is an illusion created by an "evil genius" bent on deceiving him. "I will suppose that . . . some malicious demon of the utmost power and cunning has employed all his energies in order to deceive me. I shall think that the sky, the air, the earth, colours, shapes, sounds and all external things are merely the delusions of dreams which he has devised to ensnare my judgement. I shall consider myself as not having hands or eyes, or flesh, or blood or senses, but as falsely believing that I have all these things."

That the demon and Descartes's mind were the only two realities would be the very pinnacle of deception, Descartes reasoned. Were there even one other mind as "audience" for the deception, Descartes would at least be correct about the existence of minds such as his own.

Descartes's evil genius anticipates the brains-in-vats paradox in all meaningful particulars. The Penfield experiments merely showed how Descartes's metaphysical fantasy might be physically conceivable. The illusion in the Penfield experiments was more realistic than a dream or memory, though not complete. Penfield's patients described it as a double consciousness: Even while reliving the past experience in detail, they were also aware of being on the operating table.

One can readily envision the more complete neurological illusion supposed in the brains-in-vats riddle. The eyes do not send the brain pictures, nor the ears sound. The senses communicate with the brain via electrochemical impulses in the nerve cells. Each cell in the nervous system "sees" only the impulses of neighboring cells, not the external stimulus that caused them.

If we knew more about the original sensory nerve communication with the brain (as may be the case in a century or so), it might be possible to simulate any experience artificially. That contingency throws all experience into doubt. Even the current embryonic stage of neurology is no guarantee of the validity of our senses. It might be the twenty-fifth century right now, and the forces behind the brains-in-vats laboratory want you to think it's the twentieth, when such things don't happen!

The existence of one's brain is just as open to doubt as the external world. We talk of "brains in vats" because it is a convenient picture, wryly suggestive of bad science fiction. The brain is shorthand for "mind." We no more know, with unimpeachable certainty, that our consciousness is contained in a brain than that it is contained in a body. A yet more complete version of the fantasy would have your mind hallucinating the entire world, including Penfield, J.V., and the brains-in-vats riddle.

Ambiguity

"Brains in vats" is the quintessential illustration of what philosophers call the "problem of knowledge." The point is not the remote possibility that we are brains in vats but that we may be just as deluded in ways we cannot even imagine. Few persons reach their fifteenth birthday without having some thoughts along this line. How do we know *anything* for sure?

The whole of our experience is a stream of nerve impulses. The sheen of a baroque pearl, the sound of a dial tone, and the odor of apricots are suppositions from these nerve impulses. We have all *imagined* a world that might account for the unique set of nerve impulses we have received since (and several months before) birth. The conventional picture of a real, external world is not the only possible explanation for that neural experience. We are forced to admit that an evil genius or a brains-in-vats experiment could explain the neural experience just as well. Experience is forever equivocal.

Science places great faith in the evidence of the senses. Most people are skeptical about ghosts, the Loch Ness monster, and flying saucers, not because they are inherently stupid notions, but only because no one has produced unquestionable sensory evidence for them. Brains-in-vats turns this (apparently reasonable!) skepticism inside out. How can you know, by the evidence of your senses, that you are not a brain in a vat? You can't! The belief that you are *not* a brain in a vat can never be disproven empirically. In the jargon of philosophy, it is "evidence-transcendent."

This is a serious blow to the idea that "everything can be determined scientifically." At issue is not some bit of trivia such as the color of a tyrannosaurus. If we cannot even know whether the external world exists, then there are profound limitations on knowledge. Our conventional view of things might be outrageously wrong.

Ambiguity underlies a famous analogy proposed by Albert Einstein and Leopold Infeld. In 1938 they wrote:

> In our endeavour to understand reality we are somewhat like a man trying to understand the mechanism of a closed watch. He sees the face and the moving hands, even hears its ticking, but he has no way of opening the case. If he is ingenious he may form some picture of a mechanism which could be responsible for all the things he observes, but he may never be quite sure his picture is the only one which could explain his observations. He will never be able to compare his picture with the real mechanism and he cannot even imagine the possibility of the meaning of such a comparison.

Is Anything Certain?

Descartes's evil genius was the starting point of an investigation into how we know what we know. Descartes wrote: "Some years ago I was struck by the large number of falsehoods that I had accepted as true in my childhood, and by the highly doubtful nature of the whole edifice that I had subsequently based on them. I realized that it was necessary, once in the course of my life, to demolish everything completely and start again right from the foundations if I wanted to establish anything at all in the sciences that was stable and likely to last."

Descartes wanted to address the problem of knowledge in much the same way that Euclid had treated geometry two thousand years earlier. All of Euclid's geometry is deduced from a set of five *axioms*. An axiom was, in Euclid's time, a statement so obviously true that one could not imagine a world in which it was false (for instance: "A straight line can be drawn between any two points"). All the *theorems*—provably true statements—of traditional geometry can be derived from Euclid's five axioms.

Descartes wanted to do the same thing with the facts of the real world. He needed first to identify a set of facts known with utter certainty. These facts would be the axioms of his natural philosophy. Then he would establish valid rules of inference. Finally, Descartes would use those rules to deduce new facts from the original set of incontestable facts.

Unfortunately, almost any statement about the real world has some degree of doubt. Descartes found the ground floor of his natural philosophy vanishing beneath his feet: "So serious are the doubts into which I have been thrown as a result of yesterday's meditation that I can neither put them out of my mind nor see any way of resolving them. It feels as if I have fallen unexpectedly into a deep whirlpool which tumbles me around so that I can neither stand on the bottom nor swim up to the top."

This dizzying "whirlpool" aptly describes *ontology*, the study of what is most real. The first thing to realize in constructing an ontology is that the accepted, everyday "facts" of the external world are disputable. You can almost always think of a scenario in which unquestioned beliefs could be wrong. Is Paris the capital of France? Very likely it is, but still, there is a sliver of doubt that never vanishes. It's barely conceivable that our government is a totalitarian conspiracy that, for reasons of its own, does not want its citizens to know the *real* capital of France. They've rewritten all the history and geography books, and force teachers to indoctrinate each new generation of children with the Paris fiction. You say you went to Paris last summer and saw a bunch of official-looking French government buildings there? That could have been a theme-park simulation maintained by our government to give its citizens the illusion of freedom of travel.

Wild fancies like this should not obscure the fact that some things are more disputable than others. By most people's standards, the Loch Ness monster is less real than a tyrannosaurus, and both are less real than the elephant you saw at the zoo last Sunday. What is most certain of all?

A popular answer is the truths of logic and mathematics. Even if your first-grade teacher was the dupe of a conspiracy bent on teaching you falsehoods, you cannot doubt that 2 + 2 is 4. Right now you can picture two things, and put two more things beside them, and see that the total is four. This deduction seems obviously true in any possible world, be it the external world we believe exists, the brain-in-vats laboratory, or something stranger yet.

There are two problems with this answer. First of all, you can take the ultra-skeptical position that even logic and mathematics are an illusion. Just because you don't see how you could be wrong about 2 + 2 = 4 doesn't guarantee that it's right.

Your brain is evidently in a certain state when you come to a valid conclusion of logic or mathematics. What's to stop the brains-in-vats overlords from deluding you about arithmetic as well as the physical world? Possibly 2 + 2 is 62,987, but by stimulating your brain in a precise way, the mad scientists have you thinking it's 4, and even thinking that it's *obvious* it's 4 and you can *prove* it's 4. Maybe somewhere they've got a whole row of brains in vats, each one believing in a different sum for 2 + 2 and immersed in a different "reality" consistent with that sum.

Philosophers rarely take skepticism that far. There are enough other things to doubt in the world. The other, more pragmatic problem with restricting certainty to logic and mathematics is that it leaves us with no way of justifying beliefs about the physical world. Certainty about arithmetic is not going to tell us what the capital of France is. So, are there any facts, aside from logic and mathematics, about which we can be certain?

Descartes had some interesting ruminations on this point. He noted that there are limits to imagination, possibly including that of evil geniuses. The fantastic objects of a dream or a surreal painting are based on real objects. "For even when painters try to create sirens and satyrs with the most extraordinary bodies, they cannot give them natures which are new in all respects; they simply jumble up the limbs of different animals," wrote Descartes. (Are there *any* mythical beasts which aren't simple pastiches of nature? Centaurs, minotaurs, unicorns, griffons, chimeras, sphinxes, manticores, and the like don't speak well for the human imagination. None is as novel as a kangaroo or starfish.)

Descartes would probably assert that the beings running the brains-in-vats laboratory couldn't have dreamed up *everything* from scratch. There would be, let us say, eyes and fur out there in the "real" external world outside the laboratory, even if they were not arranged like a dog. Descartes also remarked that the colors of the most fanciful paintings are *real* colors no less. For this reason, he felt justified in believing that the color red exists, even if he was deceived by an evil genius. (Do you agree? Or is it conceivable that the "real" world is black and white, and that color is a neurological illusion created by a remarkably inventive brains-in-vats research department?)

In saying that colors are real, Descartes meant the subjective sensations of color, rather than pigments, wavelengths of light, or anything else that we have come to associate with those primal sensations. In fact, Descartes concluded that the one thing one can be certain of is subjective feelings—specifically, one's own subjective feelings (for who can be sure that others think and feel as one's self?).

Suppose you doubt that your own mind exists. Then you doubt that you are doubting—which is to say, you *are* doubting after all. Something must be doing the doubting. You may be deluded in many ways, but there must at least be a mind that is being deluded. Hence Descartes's famous conclusion: "I think, therefore I am."

Idealism is the belief that only mind is real or knowable. Though not properly an idealist, Descartes inspired the movement. An idealist says that when you eat a chili pepper and burn your mouth, the *sensations* of pain or heat are indisputably real. The chili pepper itself may be an illusion: a marzipan fake doctored with Tabasco sauce, or part of a bad dream brought on by indigestion. Because pain and flavor are purely subjective, the fact of the pain or the flavor is beyond dispute. Subjective feelings transcend the physical reality of their cause.

Another example: Almost everyone has been frightened by horror movies, horror novels, and nightmares. Although it's only a movie/story/dream, the momentary fear is *real* fear. Penfield's patient J.V. was genuinely frightened by the man with the bag of snakes, even if he was (in neurological replay on the operating table) an illusion. Likewise, one cannot doubt that one is happy, sad, in love, in grief, amused, or jealous, if such mental states apply.

Subjective feelings are a very limited basis for reasoning about the external world. Descartes nevertheless believed he could deduce many significant conclusions from the fact of his own mind. From "I am," he concluded that "God exists." Every effect must have a cause, Descartes reasoned, and thus he must have a creator. From "God exists" Descartes jumped to "The external world exists" because, as a perfect being, God could not deceive us into believing in an illusory external world: He would not permit an evil genius.

Few modern philosophers accept this chain of reasoning. All things may seem to have a cause, but do we know this with total certainty? Again, cause and effect could be a fiction put in our minds by an evil genius.

Even allowing that there is a cause for one's existence, it is misleading to call that cause "God." "God" means a lot more than a cause for one's existence. Perhaps Darwinian evolution is a cause for our existence, but that is not what most people mean by "God." And even allowing that God exists, how do we know that He wouldn't countenance an evil genius?

None of this means that Descartes was wrong, but only that he was not true to the spirit of his original skepticism. One of Descartes's severest critics was Scottish philosopher and historian David Hume (1711–76). At the height of his renown, Hume was a celebrity in London and Paris but could not teach at any university because of his outspoken atheism. For a time he made a difficult living as private tutor to the Third Marquess of Annandale, who was insane. Hume doubted every step of Descartes's argument, even the existence of one's own mind. Hume said that when he introspected, he always "stumbled on" ideas and sensations. Never did he find a self distinct from those thoughts.

Hume argued that there are only two types of revealed truth. There are "truths of reason," such as $2 + 2 = 4$. Then there are "matters of fact," such as "The raven in the aviary of the Copenhagen zoo is black." This double-pronged conception of truth is called "Hume's fork." A question not of either type (such as "Does the external world really exist?") is unanswerable and meaningless, maintained Hume.

Deduction and Induction

To derive useful conclusions about the real world, we must work from premises that are (in the exacting sense of the philosophical skeptic) uncertain. Science and common sense are forever building structures of belief on uncertain foundations. No scientific conclusion is utterly certain.

There are two ways by which we know (or think we know) things, and they are closely related to Hume's distinction. One way of knowing is deduction, the "logical" way of drawing conclusions from given facts. An example of deductive reasoning is:

> All men are mortal.
> Socrates is a man.
> Therefore Socrates is mortal.

The first two lines are premises, facts assumed to be given. Deduction is the act of deriving the third line from the two lines above it. Valid deductions are truths of reason in Hume's terminology.

Descartes wanted to use deduction to derive new facts from certain premises. The new facts would then be equally certain. Fortunately, deduction may also be applied to less than certain premises. A hard-core skeptic can insist that neither of the two premises above is certain. There may be immortal men somewhere, and Socrates could have been a being from another planet. These uncertainties are transmitted to the conclusion. The deduction itself, however, is as certain as any statement of logic can be. *Whenever* "all A are B" and "C is A," it follows that "C is B." You can just as well conclude:

> All bankers are rich.
> Rockefeller is a banker.
> Therefore Rockefeller is rich.

or:

> All ravens are black.
> The bird in Edgar Allan Poe's "The Raven" is a raven.
> Therefore the bird in Edgar Allan Poe's "The Raven" is black.

This type of reasoning is called a *syllogism.* An irony of deduction is that the "subject matter" of the premises makes no difference to the deductive process. Socrates, Rockefeller, or Poe's raven, it's all the same.

The other basic way of knowing is induction. Induction is the familiar process by which we form generalizations. You see a raven. It's black. You see other ravens, and they're black too. Never do you see a raven that isn't black. It is inductive reasoning to conclude that "all ravens are black."

Both science and common sense are founded on induction. Despite his renown for "deduction," most of Sherlock Holmes's reasoning is more induction than deduction. Induction is reasoning from "circumstantial evidence" or Hume's "matters of fact." It extrapolates from observations that are not understood on a deeper level. You don't know *why* all the ravens seen have been black. Even after seeing 100,000 ravens, all black, the 100,001st raven just might be white. A white raven isn't inherently absurd, like a triangle with four sides. There is no logical necessity to an inductive conclusion.

For this reason, induction has always seemed less legitimate than deduction. Hume for one was skeptical of it. As he complained, we use inductive reasoning to justify inductive reasoning. ("Induction has stood the test of time. Therefore it should be reliable in the future.") Philosopher Morris Cohen quipped that books on logic are divided into two parts: a part one, on deduction, in which fallacies are explained, and a part two, on induction, in which fallacies are committed. (Note the modified plan of this book!)

Induction is working backward, like solving a maze by backtracking from the goal. Instead of taking a general law ("All ravens are black") and applying it to specific cases ("This bird is a raven;

therefore this bird is black"), induction goes from specific cases to a general law. Induction is founded on the belief—the hope—that the world is not essentially deceptive. From the fact that every raven ever examined was black, we conclude that *all* ravens are black, even the ones no one has ever seen. We assume that the unobserved ravens are similar to the observed ravens, that the seeming regularities of the world are genuine.

It could be that the world teems with unseen white ravens, forever behind your head, never straying into sight. Lingering uncertainty plagues every inductive conclusion. Why do we bother with inductive reasoning, then? We use it because it is the only way of getting broadly applicable facts about the real world. Without it, we would have only our trillions of experiences, each as separate and meaningless as a bit of confetti.

Induction provides the fundamental facts from which we reason about the world. Empirically tested generalizations take the place in science that Descartes hoped certain axioms would in his philosophy. The teaming of induction and deduction is the basis of the scientific method.

Confirmation Theory

The problem of knowledge has intrigued many of the keenest minds of philosophy, science, and even literature for as far back as we have records. Philosophers call this study *epistemology*. A newer term, applied in more strictly scientific contexts, is *confirmation theory*. Each is a study of how we know what we know; an investigation into the business of drawing valid conclusions from evidence.

Investigating the very process of knowing is different from investigating butterflies, nebulae, or anything else. Confirmation theory is largely a study of logic puzzles and paradoxes. To the uninitiated, this probably sounds as peculiar as a study based on mirages. By their nature, paradoxes expose the cracks in our structures of belief. Bertrand Russell said, "A logical theory may be tested by its capacity for dealing with puzzles, and it is a wholesome plan, in thinking about logic, to stock the mind with as many puzzles as possible, since these serve much the same purpose as is served by experiments in physical science."

The past few decades have been a very fruitful time for paradoxes of knowledge. This book discusses a selection of recent paradoxes that are so significant and mind-bending that they deserve a place in the mental bestiary of any broadly educated person.

Paradox

It is best to start by explaining what is meant by a paradox. The word is used in different ways, of course, but at the heart of all the uses is contradiction. A paradox starts with a set of reasonable premises. From these premises, it deduces a conclusion that undermines the premises. It is a travesty of the notion of proof.

One thing not immediately apparent (if the paradox is clever enough) is the reason for the contradiction. Is it possible for a perfectly valid argument to lead to contradiction, or is that "guaranteed" not to happen?

Paradoxes can be loosely classified according to how and where (if anywhere) the contradiction arises. The weakest type of paradox is the fallacy. This is a contradiction that arises through a trivial but well-camouflaged mistake in reasoning. We've all seen those algebraic "proofs" that 2 equals 1, or some other absurdity. Most are based on tricking you into dividing by 0. One example:

1. Let $x = 1$
2. Then obviously: $x = x$
3. Square both sides: $x^2 = x^2$
4. Subtract x^2 from both sides: $x^2 - x^2 = x^2 - x^2$
5. Factor both sides: $x(x - x) = (x + x)(x - x)$
6. Factor out the $(x - x)$: $x = (x + x)$
7. Or: $x = 2x$
8. And since $x = 1$: $1 = 2$

The fatal step is dividing by $(x - x)$, which is 0. Line 5, $x(x - x) = (x + x)(x - x)$, correctly asserts that 1 times 0 equals 2 times 0. It does not then follow that 1 equals 2; any number at all times 0 equals any other number times 0.

In a fallacy, the paradox is an illusion. Once you spot the error, all is right with the world again. It might seem that all paradoxes are like this deep down. The error may not be as obvious as in the example above, but it is there. Rout it, and the paradox vanishes.

Were that all there is to paradox, confirmation theory and epistemology would be simpler and less interesting fields. We will not be concerned with simple fallacies. Many paradoxes are valid and disturbing.

More powerful paradoxes often take the form of a *thought experiment* (sometimes known by the German name *Gedankenexperiment*). This is a situation that may be imagined but which (usually) would be difficult to carry out in practice. Thought experiments typically show that certain conventional assumptions can lead to an absurdity.

One of the simplest and most successful of thought experiments was that devised by Galileo to demonstrate that heavy objects do not fall faster than light ones. Suppose (as was the belief in Galileo's day) that a 10-pound lead ball falls faster than a 1-pound wooden ball. Imagine that you connect the balls with a string and drop them from a great height. The wooden ball, being lighter, will lag behind the lead ball, pulling the string taut. Once that happens, you have a wooden ball weighted down by a lead ball: an 11-pound system that, being heavier yet, ought to fall faster than either ball alone. Does the system speed up once the string is taut? Although not strictly impossible, that conclusion is suspect enough to cast doubt on the original assumptions. Unlike most thought experiments, Galileo's was easily carried out. Galileo dropped objects of different weight (but not, as story has it, from the leaning Tower of Pisa) and found that they fell at the same rate. The uniformity of gravitational acceleration has been so well accepted, in fact, that we see nothing paradoxical about Galileo's thought experiment today.

The sense of paradox is sharper in another famous thought experiment, the "twin paradox." The theory of relativity claims that time passes at different rates according to the motion of the observer. Let one of a set of identical twins blast off in a rocket, travel at near the speed of light to Sirius, and return to Earth. According to relativity, he will find he is years younger than his twin brother. He will be younger by his calendar watch, by the number of wrinkles and gray hairs, by his subjective impression of how much time has elapsed, and by any other physically meaningful definition of time we know of.

When first formulated, the twin paradox was so contrary to experience that many (including French philosopher Henri Bergson) cited it as proof that relativity must be wrong. Nothing in everyday life leads us to believe that time is relative. A pair of twins are the same age from the cradle to the grave.

Today, the twin paradox is an accepted fact. It has been tested in numerous experiments—not with twins, but with extremely accurate clocks. In a 1972 experiment designed by physicist Joseph Hafele, cesium clocks transported around the globe on commercial jetliners established that the human passengers returned home a minuscule but measurable split second younger than everyone else. No physicist doubts that if an astronaut did travel at near the speed of light, he would return younger than a stay-at-home originally the same age.

The paradox lies in our mistaken assumptions about the way the world works rather than in the logic of the situation. The unspoken premise of the twin paradox is that time is universal. The twin paradox demonstrates that this premise is untenable: Common sense is wrong. You might not think there is an animal that has fur and lays eggs, but here's the platypus, a living paradox—sort of. There is, of course, no logical necessity that a fur-bearing animal not lay eggs, nor one that time not depend on the motion of the observer.

This, then, is the second type of paradox, the "common sense is wrong" type. In these paradoxes, the contradiction, while surprising, can be resolved. It is fairly obvious which of the original assumptions must be discarded, and however painful it may be to relinquish that assumption, once it is thrown out, the contradiction vanishes.

There are stronger paradoxes yet. Neither the fallacy nor the "common sense is wrong" variety has the tantalizing quality of the best paradoxes. These most paradoxical of paradoxes defy resolution.

A very simple example of genuine paradox is the "liar paradox." Devised by Eubulides, a Greek philosopher of the fourth century B.C., the paradox is often wrongly attributed to Epimenides, who is merely the fictionalized speaker (as is Socrates in Plato's dialogues). Epimenides of Crete allegedly said, "All Cretans are liars." To convert this into a full paradox, cheat a bit and playfully define a liar as someone whose every utterance is false. Then Epimenides would have been saying in essence, "I am lying," or "This sentence is false."

Take the latter version. Is the sentence true or false? Suppose that "This sentence is false" is true. Then the sentence is false because it's a true sentence, and that's what it asserts!

All right, then, it must be false. But if "This sentence is false" is false, then it must be true. This gives us two reductio ad absurdum arguments. If it's true it's false, so it can't be true, and if it's false it's true, so it can't be false. The paradox is intrinsic and indelible.

With this third type of paradox, it is not at all clear which premise should be (or *can* be) discarded. These paradoxes remain open questions. The paradoxes to be discussed in this book are at least of the second type, and mostly of the third. Be warned that few have a universally accepted solution.

The best paradoxes raise questions about what kinds of contradictions can occur—what species of impossibilities are possible. Argentine writer Jorge Luis Borges (1899–1986), whose work appeals to all lovers of paradox, explored many such questions in his short stories. In "Tlön, Uqbar, Orbis Tertius," he describes an encyclopedia, supposedly from another world, created as an elaborate hoax by a group of scholars. Borges's scholars even imagine the paradoxes of their fictitious world; so alien is the thinking of "Tlön" that their paradoxes are commonplaces to us. The greatest paradox of Tlön is that of the "nine copper coins":

> On Tuesday, X crosses a deserted road and loses nine copper coins. On Thursday, Y finds in the road four coins, somewhat rusted by Wednesday's rain. On Friday, Z discovers three coins in the road. On Friday morning, X finds two coins in the corridor of his house. . . . The language of Tlön resists the formulation of this paradox; most people did not even understand it. The defenders of common sense at first did no more than negate the veracity of the anecdote. They repeated that it was a verbal fallacy, based on the rash application of two neologisms not authorized by usage and alien to all rigorous thought: the verbs "find" and "lose," which beg the question, because they presuppose the identity of the first and of the last of the nine coins. They recalled that all nouns (man, coin, Thursday, Wednesday, rain) have only a metaphorical value. They denounced the treacherous circumstance "somewhat rusted by Wednesday's rain," which presupposes what is trying to be demonstrated: the persistence of the four coins from Tuesday to Thursday. They explained that *equality* is one thing and *identity* another, and formulated a kind of *reductio ad absurdum:* the hypothetical case of nine men who on successive nights suffer a severe pain. Would it not be ridiculous—they questioned—to pretend that this pain is one and the same? . . . Unbelievably, these refutations were not definitive. . . .

To the Tlön way of thinking, the "nine copper coins" has the quality of true paradoxes, that it is never fully explained away. It is interesting to wonder if our paradoxes would seem as banal to the inhabitants of another world. Are paradoxes "all in our heads" or are they built into the universal structure of logic?

Science as a Map

This book deals with paradoxes of knowledge; paradoxes that illuminate how we know things. At first sight, the idea of knowing what the universe is like is absurd. Penfield's experiments demonstrated that memories occupy *engrams,* specific physical sites in the brain. To know about Chief Crazy Horse or frost or Tasmania is to have some part of your brain that represents Chief Crazy Horse, frost, or Tasmania. These brain sites may wander and interpenetrate, and the whole story about how memories are stored and recalled is probably a lot more complicated than we can imagine today. That granted, engrams are not infinitely small. Your mental representation of Chief Crazy Horse occupies a part of your brain's storage capacity that cannot be occupied simultaneously by anything else.

You might naïvely picture the brain as containing scale models of things in the outside world. It is evident that these models must leave out much detail. The very fact that the universe is so much larger than your head makes universal knowledge unattainable. There is no way a human brain can contain representations of everything in the world.

That our brains work as well as they do indicates that they are selective in what they retain. The primary tool for condensing the complexity of the world is generalization. Our brains do this at many levels. Science is a conscious and collective way of simplifying through generalization. It is a means of packing the great, vast universe into our tiny brains.

Science is a mnemonic device. Rather than remembering what has happened to every apple released from its support, we remember gravity. It is a map of the external world. Like any map, it omits detail. Small towns, trees, houses, and rocks are left out of road maps to make room for highways, coastlines, national boundaries, and other features judged more significant to the map's users. Comparable judgments face the scientist.

Science must be more than a desultory catalogue of information. It must embrace not only the collecting of information but the understanding of it. What is understanding? Surprisingly, this philosophical question can be given a rather exact, if preliminary, answer.

Paradox and SATISFIABILITY

Often it is easier to draw a boundary around an unknown than to describe it. Thomas Jefferson did not know what was in the Louisiana Territory, only its boundaries. It is convenient to take this approach in describing what it means to understand a body of information.

At a bare minimum, understanding entails being able to detect an internal contradiction: a paradox. If you cannot even tell whether a set of statements are self-contradictory, then you don't really understand the statements; you haven't thought them through. Think of the querulous schoolteacher who slips a contradiction into the lecture to see if a daydreaming student will agree with it:

"Isn't that right, Miriam?"
"Uh . . . yes, ma'am."
"I see. Someone obviously hasn't been listening to a word I've been saying."

Spotting contradiction is not all there is to understanding. There is probably much more. But it is certainly a prerequisite. By exposing a contradiction in a set of assumptions, the author of a paradox shows that we don't understand as much as we think we do.

In logic, the abstract problem of detecting paradox is called SATISFIABILITY. (This and related logic problems are usually printed in capital letters.) Given a set of premises, SATISFIABILITY asks, "Do these statements necessarily contradict?" Another way of phrasing it is, "Is there any possible world in which these premises can all be true?"

SATISFIABILITY deals in logical abstractions, not necessarily the truths of the real world. Consider these two premises:

1. All cows are purple.
2. The King of Spain is a cow.

One's natural reaction is that both statements are wrong. But something can be wrong without being a paradox. You can at least imagine a world in which both these statements are true. Logicians say a set of statements are *satisfiable* when they are true in some possible world—even if not our own.

The situation is different here:

1. All cows are purple.
2. The King of Spain is a cow.
3. The King of Spain is green.

In no possible world can all three statements be simultaneously true (assuming that colors like purple and green are mutually exclusive). There is a paradox; the statements are said to be unsatisfiable.

Notice that no single statement is to blame for the contradiction. You could strike out any one statement and have a possible state of affairs. The paradox lies in the way the three statements intertwine.

This oddity turns out to be incredibly significant. Because paradox cannot be localized, SATISFIABILITY is extremely difficult in general. It is in fact notoriously hard, a paragon of difficulty. It is difficult in the sense that as the number of premises is increased, the time required to check them for possible contradiction increases at a staggering rate. The increase is so great that many SATIS-FIABILITY problems with a hundred or more premises may be for all practical purposes insoluble. Even if these problems were turned over to the fastest computer in existence, they would take practically forever to solve.

We can use paradoxes as a metaphor, a way of marking off the limits of understanding. Science tries to find simple generalizations that account for myriad facts. Whenever we find ourselves unable even to detect blatant contradictions in a body of knowledge or belief, then we do not understand it. The difficulty of SATIS-FIABILITY is a rough measure of how difficult it is to "compress" empirical information with generalizations. SATISFIABILITY sets a loose limit on the difficulty of acquiring information and deducing conclusions from it.

The Universal Problem

The early 1970s saw a surprising discovery in mathematical logic. Two seminal papers by computer scientists Stephen Cook (1971) and Richard Karp (1972) revealed that many types of abstract logic problems are really the *same* problem in disguise. They are all equivalent to SATISFIABILITY, the problem of recognizing paradox.

The class of problems equivalent to SATISFIABILITY is called "NP-complete" (don't worry about the name for now). One surprising thing about the NP-complete problems is how (apparently) dissimilar they are. Karp's paper listed twenty-one NP-complete problems, including the "traveling-salesman" problem (an old mathematical riddle) and the "Hamiltonian Circuit" problem, based on a puzzle novelty that was a faddish nineteenth-century predecessor of Rubik's cube. Over the years, the list of problems known to be NP-complete has grown enormously.

Problems of finding your way through a labyrinth, of decoding ciphers, and of constructing crossword puzzles are NP-complete. The NP-complete problems include generalized versions of many classic logic puzzles or brainteasers: the sort of recreational logic most associated with Martin Gardner and Raymond Smullyan in recent years, and Sam Lloyd, Lewis Carroll, Henry Ernest

Dudeney, and many others, known and anonymous, before them. That such diverse problems can be essentially the same was wildly unexpected. It is not too much of a hyperbole to compare Cook and Karp's discoveries to the discovery that everything is made of atoms. Much of the intellectual difficulty of the world, both profound and frivolous, is made of the same stuff. NP-completeness is a cosmic riddle; a paradigm of the inscrutability of a universe of exponentially vast possibilities to a finite mind.

When logicians say that all NP-complete problems are, in effect, the same problem, they mean that an efficient general solution to *any* NP-complete problem could be transformed in such a way as to solve all the other problems. If ever anyone solved one NP-complete problem, then *all* the NP-complete problems would melt away like cotton candy in a summer rain.

It is as if you discovered that all the famed treasures of the world can be unlocked with the same key—*if* that key exists. Is there an efficient solution to any/all NP-complete problems? This is one of the deepest unresolved questions in mathematical logic today.

Paradox is thus a much deeper and more universal concept than the ancients would have dreamed. Rather than an oddity, it is a mainstay of the philosophy of science. Paradoxes are both appealing and haunting. There is a subversive joy in seeing logic tumble like a house of cards. All the well-known paradoxes of confirmation theory and epistemology were conceived more or less in the spirit of intellectual play. In few other fields is it possible for the interested nonexpert to sample so much of the true flavor of the field and have fun doing it. How we know—the interplay of induction and deduction, of ambiguity and certainty—is the theme of the paradoxes to follow.

2

INDUCTION

Hempel's Raven

THE BEST-KNOWN modern paradox of confirmation was proposed by German-born American philosopher Carl G. Hempel in 1946. Hempel's "paradox of the ravens" deals with induction, the drawing of generalizations. It is a mischievous reaction to those who think that science may be resolved into a cookbook scientific method.

Hempel imagined a birdwatcher trying to test the hypothesis "All ravens are black."[1] The conventional way of testing that theory is to

---

[1] Ornithological note: "Raven" usually means a single species, *Corvus corax*, found worldwide in the Northern Hemisphere. This is the raven of Poe's poem. Ravens are iridescent black with glints of green, purple, and blue predominating. Mexico and the American Southwest also have a smaller bird called the Chihuahuan raven *(Corvus cryptoleucus)*. This bird is black with a white neck that is exposed when the bird crooks its head. I have failed to find

seek out ravens and check their color. Every black raven found confirms (provides evidence for) the hypothesis. On the other hand, a single raven of any color other than black disproves the hypothesis on the spot. Find even one red raven and you need look no further: The hypothesis is wrong.

All are agreed on the above. Hempel's paradox begins with the claim that the hypothesis may be restated as "All nonblack things are nonravens." Logic tells us that this is entirely equivalent to the original hypothesis. If all ravens are black, then certainly anything which isn't black can't be a raven. This rewording is known as a *contrapositive*, and the contrapositive of any statement is identical in meaning.

"All nonblack things are nonravens" is a lot easier to test. Every time you see something that *isn't* black, and it turns out *not* to be a raven, this restated hypothesis is confirmed. Instead of looking for ravens on damp, inaccessible moors, you need only look for nonblack things that aren't ravens.

A blue jay is sighted. It's nonblack and it's not a raven. That confirms the contrapositive version of the hypothesis. So does a pink flamingo, a purple martin, and a green peacock. Of course, a nonblack thing doesn't even have to be a bird. A red herring, a gold ring, a blue lawn elf, and the white paper of this page also confirm the hypothesis. The birdwatcher does not have to stir from his easy chair to gather evidence that all ravens are black. Wherever you are right now, your visual field is filled with things that confirm "All ravens are black."

Now clearly this is ridiculous. There is yet a further absurdity. To see it, suppose that you want to dismiss the paradox by saying, all right, evidently the blue jay or the red herring *does* confirm "All ravens are black" to some infinitesimal degree. If you could summon up a magic genie, capable of examining all the nonblack things in the world in the blink of an eye, and if that genie found that not one of those nonblack things was a raven, that assuredly would prove that there are no nonblack ravens—that all ravens are black. Maybe it is not so incredible that a red herring could confirm the hypothesis.

Don't get too cozy with this resolution. It is easy to see that that

any mention of albino or other distinctly nonblack ravens, but would not be surprised to learn that such birds exist.

None of this, of course, has anything to do with the present discussion. Aside from this footnote, I will assume that the color of ravens is perfectly well defined, and that no one has ever seen a raven that is any color other than black.

same red herring *also* confirms "All ravens are white." The contrapositive of the latter is "All nonwhite things are nonravens," and the herring, being a nonwhite thing, confirms it. An observation cannot confirm two mutually exclusive hypotheses. Once you admit such a patent contradiction, it is possible to "prove" anything. The red herring confirms that the color of all ravens is black, and also that that color is white; ergo:

Black is white. QED.

Reasonable assumptions have led to resounding contradiction.

To scientists, Hempel's paradox is more than a puzzle. Any hypothesis has a contrapositive, and confirming instances of the contrapositive are often very easy to find. Something is certainly wrong. But what?

Hempel's raven is a good introduction to the perils and puzzles of confirmation. Of all the major paradoxes to be discussed, it is among the most nearly resolved. It will be worthwhile to back up a bit before coming to the resolution, though, to discuss the background of the paradox.

Confirmation

To put it in as few words as possible, confirmation is the search for truth. It is the mainspring of science, and more than that, it is something we do every day of our lives.

Analyzing confirmation is almost like analyzing sneezing: We know what it is, but it is usually so automatic that it is hard to say exactly how it is done. The paradoxes of confirmation probably owe a lot to this shared set of subconscious expectations. These expectations can lead us astray.

As you remember from high school, there is a "scientific method" that goes roughly as follows. You form a hypothesis—a guess about how the world works. Then you try to test it through observation or experiment. The evidence you gather either confirms the hypothesis or refutes it. Like much of what you learn in high school, this is true while leaving important things unsaid.

Most useful hypotheses are generalizations. Hempel's paradox plays off a bit of common sense called "Nicod's criterion" after philosopher Jean Nicod. To put it in terms of black ravens, this says that (a) sighting a black raven makes the generalization "All ravens are black" more likely; (b) sighting a nonblack raven disproves the statement; and (c) observations of black nonravens and nonblack nonravens are irrelevant. A black bowling ball or a blue lawn elf cannot tell us anything about the color of ravens. Nicod's criterion is behind all scientific inquiry, and if something is wrong with it, we are in serious trouble indeed.

The sighting of a black raven furnishes evidence in favor of the hypothesis that all ravens are black, but of course does not *prove* that the hypothesis is right. No single observation can do that. Sightings of black ravens, in the absence of ravens of any other color, increase your confidence (reasonably enough) that all ravens are black.

Confirmation is trickier than it appears. You might think that the more confirming evidence for a hypothesis, the more likely it is to be true. Not necessarily. It is possible for two confirming observations to prove a hypothesis *false*. That is what happens in the following thought experiment, inspired by philosopher Wesley Salmon.

Matter and Antimatter

Suppose that some of the planets in the universe are made of matter and some of antimatter (as has been speculated). Matter and antimatter look exactly alike. There is no way of telling, by examining a distant star in a telescope, whether it is made of matter or antimatter. Even the star's light gives nothing away, for the photon is its own antiparticle, and an antimatter star shines with the same kind of light as a regular star. The only thing is, when antimatter touches regular matter—BOOM!!! Both are annihilated in a tremendous explosion.

This unfortunate fact makes interstellar contact hazardous. A spaceship from planet X chances upon a spaceship from planet Y. They radio messages to each other (radio waves are made of photons too and are neither matter nor antimatter). Computers on board the ships decipher the alien languages and establish diplomatic relations. The two spaceships agree to dock and exchange goodwill ambassadors. Everything is fine until the last moment. Then the rockets make contact and BOOM!!!—or not, depending on the composition of planets X and Y. Whenever one is matter and the other antimatter, both ships are blown to smithereens. (There's no explosion if *both* spaceships are antimatter.)

One day, astronomers here on Earth report that they've sighted two tiny points of light that may be spaceships approaching each other. They're not sure the objects are spaceships, but on the basis of past experience the astronomers can say that for each point of light there is a 30 percent chance it is a spaceship and a 70 percent chance it is an irrelevant natural phenomenon. It is also known from past experience that any pair of spaceships that approach closely always do dock. All the other species in the galaxy seem oblivious to the matter/antimatter problem, and have to learn the hard way.

So the big question is: Will they blow up or not? Oddsmakers in Las Vegas start accepting ghoulish bets on whether there will be an annihilation. The oddsmakers reason like this: It is known that two-thirds of the planets in the universe are made of matter and one-third are antimatter. Thus for each point of light there is a 70 percent chance it is a natural phenomenon of no interest here; a 20 percent chance it is a spaceship made of matter; and a 10 percent chance that it is an antimatter spaceship.

Call the two points of light A and B. An annihilation can occur in one of two mutually exclusive ways. Either object A is a matter spaceship and B is an antimatter spaceship, or A is an antimatter spaceship and B is a matter spaceship. The chance of the first case is 20 percent of 10 percent, or 2 percent. The odds of the second case is 10 percent of 20 percent; again, 2 percent. Since these two possibilities are mutually exclusive, the total chance of an annihilation is 2 percent plus 2 percent: 4 percent.

The oddsmakers set the payoffs to bettors based on this calculated probability. Now suppose that a space prospector, returning home to Earth, grazes object A in a trillion-to-one freak accident. The prospector learns that object A *is* a spaceship and is made of ordinary matter (this from the fact that there was no explosion). Arriving on Earth, the prospector finds out about the possible annihilation and the Las Vegas book on it.

The prospector would do well to exploit his "inside information" and bet on annihilation. He knows for a fact that object A is a spaceship, whereas everyone else thinks it is probably (70 percent chance) just an asteroid or some other natural body. Given that A is a spaceship of ordinary matter, there is a 10 percent chance of annihilation, since that is the chance that object B is a spaceship and made of antimatter. The oddsmakers have set the chance at 4 percent, but the prospector, with his more complete knowledge, can set the chance at 10 percent.

Fine. Now what if another space prospector had an identical accident with object B, and also determined that it is a spaceship made of matter? This second prospector could of course use the same reasoning to arrive at the same conclusion: that the chance of annihilation has been boosted from 4 percent to 10 percent. *But* the combined information of the two prospectors actually rules out an annihilation entirely. They have determined that the spaceships are both made of the same kind of matter as Earth, and that means that the chance of annihilation is a big fat zero!

Absolute and Incremental Confirmation

Two items of evidence (the prospectors' collisions with the spaceships) each confirm the hypothesis that there will be an annihilation, even though the observations together refute it. I prefer to call this an irony rather than a paradox, for there is no doubt that such strange turns of affairs can exist. The probability calculations of the oddsmakers, of the prospectors, and of us, aware of both prospectors' experiences, are sound. These peculiar situations have been studied intensively by confirmation theorists.

The peculiarity is partly semantic. The verb "confirm" is used in two ways. In everyday speech, we almost always use "confirm" in the *absolute* sense, to mean that something is clinched; established beyond reasonable doubt. "The boss confirmed that Sandra got the raise" means that, whatever the doubts beforehand, it is now just about 100 percent certain that Sandra got the raise.

Hardly any experiment provides absolute confirmation of a hypothesis. Scientists and confirmation theorists often use "confirm" in the *incremental* sense. To confirm incrementally is to "provide evidence for" or "increase the probability of." We speak of probabilities because confirmation of a generalization is always tentative.

You can incrementally confirm a hypothesis that was, and still is, unlikely to be true. We would not say "The boss confirmed that Sandra got the raise" to mean that some equivocal comment of the boss's has upped the chance of Sandra's getting the raise from 15 percent to 18 percent. But that type of confirmation is typical of scientific research.

Incremental confirmation is common to ironic situations such as the spaceship annihilations. Each prospectors' information increments a low chance of annihilation (4 percent) to a greater but still low chance (10 percent). Taken together, their information decreases the chance to zero. It is reassuring that such flukes vanish when the chances are higher, when a hypothesis is close to being confirmed in the absolute sense.

You can demonstrate this by playing with the odds a little. Recast the situation, giving the oddsmakers a better handle on the real state of affairs. For each object, the odds are now 10 percent that it is a natural phenomenon, 80 percent that is a matter spaceship, and 10 percent that it is an antimatter spaceship. Then the oddsmakers must set the chance of annihilation at (80 percent of 10 percent) plus (10 percent of 80 percent)—or 16 percent. Each prospector, upon learning for certain that one of the objects is a matter spaceship, can figure (as before) that the chance of annihilation is 10 percent, the chance that the other object is an antimatter spaceship. Now each prospector's estimate is *less* than the oddsmakers'. This is as it should be, since they know more than the oddsmakers, and the actual chance happens to be zero.

Counterexamples

As this shows, confirmation is only half the story. Evidence may also refute or disconfirm a hypothesis. Philosophers of science, notably Sir Karl Popper, emphasize the role of refutation.

You might think that it's just a matter of saying a glass is half full or half empty. Actually, there is an asymmetry between confirmation and refutation. It is much easier to refute a generalization than to prove it.

A counterexample is an exception to a putative rule. A white raven is a counterexample to the hypothesis that all ravens are black. A white raven does not merely make the hypothesis less likely. It proves the hypothesis wrong in one fell swoop. Logicians call this *modus tollens,* or "denying the consequent."

Rarely is the situation so simple in practice. There have been many "counterexamples" to the hypothesis that there is no Loch Ness monster. Every alleged sighting is one. Yet most scientists continue to believe that there is no Loch Ness monster. It is evident that not all supposed counterexamples have enough weight to refute an otherwise confirmed hypothesis.

Most hypotheses on the edges of current knowledge can be tested only in situations where many "auxiliary" hypotheses are tested as well. Auxiliary hypotheses are background assumptions about how the main hypothesis fits into the general body of knowledge; how microscopes, telescopes, and other equipment necessary to test the hypothesis operate; and so on. These auxiliary hypotheses often rule out any quick use of *modus tollens.*

Wesley Salmon cited a neat case of two similar counterexamples leading to rejection of auxiliary and main hypotheses, respectively. Newton's theory of gravity makes predictions about the future motions of the planets. In the nineteenth century, these predictions for the orbit of Uranus were found to be slightly, but consistently, wrong.

Some astronomers wondered if the discrepancies might be due to an unknown planet beyond Uranus. Once this planet (Neptune) was discovered in 1846, Newton's theory was not only removed from doubt but strengthened. Neptune was further evidence for Newton's theory.

At about the same time, other irregularities were noted in Mercury's orbit. Astronomers also tried to find a planet near Mercury that might account for the deviation. French amateur astronomer D. Lescarbault reported seeing a planet within Mercury's orbit in 1859. The planet was accepted as real and named Vulcan by Urbain Jean Leverrier, co-discoverer of Neptune. Subsequent astronomers could not find the planet, though, and it was soon branded a mistake. Mercury continued to depart from its predicted orbit. The deviations were not haphazard but regular, and distinctly different from what Kepler's laws (founded on Newton's gravity) predicted.

In this case, the discrepancies were ultimately accepted as evidence that Newton's theory of gravity is wrong. Mercury's wobbling orbit was one of the earliest confirmations of Einstein's general relativity.

The history of Neptune and Vulcan demonstrates two features of counterexamples. First, a counterexample may refute an auxiliary hypothesis rather than the main one. It is important to find out which is at fault. There is usually such ample room for speculation that instant refutations are rare. Second, when a theory is thrown out, it is in favor of a broader theory that makes many of the same predictions as the original. Under typical conditions in the solar system, Einstein's general relativity predicts gravitational effects all but identical to those of Newton's simpler theory. The difference turns up only in very intense gravitational fields. Of the planets, Mercury, being closest to the sun, is most subject to these relativistic effects. It alone seems to be out of step with Newton's laws.

Crank Theories

Not only should a new theory account for the successful predictions of the theory it would replace. It should offer new, different predictions of its own. In Karl Popper's terms, the new theory must have greater "empirical content." It must make more testable predictions in more realms of experience than the old theory.

A new theory should be *more* open to possible refutation, not less. If there is one thing that is a dead giveaway for a crank theory, it is that the theory has been modified to restrict its own refutation. An honest hypothesis is open to being disproven. It's one thing to say, there's a ghost that appears in the old Miller mansion at the crack of midnight whenever there's a full moon. That kind of hypothesis is worthy of attention provided there is any reasonable evidence to support it: say, testimony of a few reliable eyewitnesses. Far more typical are ghost stories that restrict refutation: A ghost appears, but never when skeptics are around.

These restrictions usually indicate that a hypothesis has failed the first stages of the confirmation process and is being kept alive by those who wish to believe it regardless of its truth. No one started out believing that

• channelers have such erratic recall of their past lives that you can't expect them to know checkable historic data (like the name of the contemporary pharaoh's wife); or
• UFOs purposely abduct people who won't be believed by the "establishment" so that the aliens' presence will remain unknown; or
• bigfoot remains disintegrate with extraordinary rapidity, so no skeletons are found (or bigfoots scrupulously bury their dead, like us humans); or
• the stars (of astrology) impel, not compel.

All these provisos were tacked on after confirmation failed to materialize. That doesn't automatically mean that the modified hypotheses are false, but it is hardly encouraging. If the process of modifying to restrict refutation continues long enough, the ultimate result is the type of hypothesis that Popper sardonically calls "irrefutable." This may sound good, but think about what it means. It is a hypothesis that cannot possibly be proven false—one so wishy-washy that no possible observation is incompatible with it. That kind of hypothesis doesn't really say anything.

The proposition that "ESP exists, but it is so iffy that even the best psychics may do no better than chance in controlled experiments"—which is essentially what some ESP apologists have said— is beyond refutation. You might ask, "How would the world be any different if ESP *didn't* exist?"

Why can't scientists give poorly supported hypotheses the benefit of the doubt? The main reason is that many, many hypotheses can be devised to account for any fixed body of data. If we say, "Okay, ESP exists, because no experiment has ruled it out" (which is true), we would have to allow a multitude of equally unrefuted hypotheses. In the end it is a desire for simplicity that leads scientists to accept only those hypotheses that can be confirmed. Indeed, says Popper, the aim of science should be to try to eliminate as many hypotheses as possible with new data.

Contrapositives

The basics of confirmation in place, let's return to Hempel's paradox with this added perspective. The first thing that concerns most people hearing the paradox for the first time is this business about the contrapositive. "Nonblack things" and "nonravens" are awkward constructions. Is "All nonblack things are nonravens" really equivalent to "All ravens are black"? If it's *not,* there is no paradox.

Here is a good way to see that they *are* logically equivalent. Forget about our human and imperfect attempts at knowledge. Pretend that we have at our service a genie who can ascertain any and all *specific* facts instantly. In other words, the genie can determine any of Hume's "matters of fact"—the direct, sensory results of any observation, without any interpretation, interpolation, or editorial comment.

Also like Hume, the genie claims that it doesn't quite understand generalizations. So if you want to know whether a statement such as "All ravens are black" is true, you have to explain it to the genie as an aggregate of individual observations. You have to tell the genie exactly what he should do to determine if Hempel's hypothesis is right or wrong.

It may come as a surprise that observations of black ravens are virtually irrelevant to the *ultimate* truth or falsity of "All ravens are black." This flatly contradicts the foregoing discussion, but remember we are now talking about the genie and not humans. The genie is going to determine the final, cosmic truth of the statement, not merely find evidence to support it. Observations of black ravens can neither prove nor disprove the statement.

Suppose the genie found a black raven. Would that prove that all ravens are black? Of course not. Suppose the genie found a million black ravens. Would that prove it? No; there could still be ravens of other colors. The statement "All swans are white" was supported by all available evidence until the discovery of Australia. There are black swans in Australia.

Suppose that the universe is infinite and there is an infinity of other planets so similar to Earth that they have black ravens on them and that the genie thereby finds an infinite number of black ravens. Would *that* prove it? No; for the same reason. At this point the genie would rightly get impatient with us, for evidently no amount of black ravens will settle anything. Looking for black ravens is a wild-goose chase.

Think about it, and you will realize that the crux of the matter is nonblack ravens. The only way Hempel's statement can be *wrong* is for there to be a raven somewhere that isn't black. The only way the statement can be *right* is for there to be no such raven. To decide ultimate truth or falsity, the genie must search for nonblack ravens. If he finds even one, the statement is irretrievably false. If he searches the entire universe—everywhere a nonblack raven could possibly be—and finds none, then Hempel's statement is unimpeachably true.

In a pragmatic sense, "All ravens are black" only *seems* to be talking about black ravens. When you translate it into an operational definition for the genie, it really says: "There is no such thing as a nonblack raven."

Now let's have the genie test the contrapositive statement, "All nonblack things are nonravens." This is another pie-in-the-sky generalization incomprehensible to the genie. We explain: "The only way 'All nonblack things are nonravens' can be wrong is for there to be at least one nonblack raven. The only way it can be right is by the complete absence of nonblack ravens everywhere."

This is just how we explained the original statement. What you must do to prove or refute "All ravens are black" is identical to what you must do to prove or refute "All nonblack things are nonravens." That is strong grounds for asserting that the two statements are equivalent.

You might object that there is one slight difference. Does not the truth of "All ravens are black" imply that there is at least one black raven?

Take the hypothesis "All centaurs are green." The genie, looking for nongreen centaurs, would find none and report the statement true. Of course, there are no centaurs of any description. It sounds funny to say the statement is true, then.

This point is again one of semantics. Logicians *do* allow that statements such as "All centaurs are green" and "If X is a centaur,

X is green" are true. For various reasons it is most convenient to do so. Hence, to a logician, there is no distinction whatsoever between a statement and its contrapositive.

You are free to dissent and insist that there must be at least one green centaur for the statement to be true. Doing so creates this slight asymmetry between Hempel's original hypothesis and the contrapositive: With the original, you have to tell the genie to make sure there is at least one black raven before reporting the statement to be true. With the contrapositive, the genie must find at least one nonblack nonraven (like a red herring). I do not think that this significantly alters the essential equivalence of the statements. Finding the obligatory black raven or red herring is but a formality; the genie's real task in either case is making sure there are no nonblack ravens.

Never Say Never

A "negative hypothesis" is one that claims that something doesn't exist. It is extremely difficult to prove negative hypotheses. ("Never say never.") It is one thing for a genie to check every place a nonblack raven could be and thus prove that there is no such thing. It is something else for us humans.

You set off on a raven-hunting expedition, see lots of black ravens, and don't find any nonblack ravens. At length you start to get sick of the whole business. All your friends say you'll *never* find a nonblack raven. When is it okay to call it quits and stop looking?

As a practical matter, you do quit sooner or later. Thereafter you feel pretty confident that there are no nonblack ravens. This does not begin to prove that all ravens are black in a logically rigorous sense, though. To do that, you would indeed have to check everywhere in the universe that a raven might be. That is obviously an unreasonable requirement.

Philosophers have a word for processes requiring an infinity of action: *supertasks*. Some philosophers think that when ascertaining something requires an infinity of actions, it cannot be known at all. Michael Dummett gave this example: "A city will never be built at the North Pole." To test this, you might hop in a time machine, set it for a given year, and travel to that year to see if a city exists at the North Pole. If not, you set the time machine for a different year, and try again. You could know whether a city will exist at the North Pole at any point in time, but knowing whether it will *ever* be built is something else again. Knowing that requires knowing an infinity of facts; doing an infinite amount of research.

If the universe is infinite, then "There are no nonblack ravens" is another proposition requiring an infinity of observations. Our genie is capable of empirical supertasks, but we are not. This is really why we confirm from sightings of black ravens and not from failures to sight nonblack ravens. The number of black ravens seen is a way of "keeping score" while actually looking for a counterexample. The more black ravens we have seen, without seeing any nonblack ones, the more confident we feel that there are no nonblack ravens. Nicod's criterion says that black ravens are a better way of keeping score in the progress of confirmation than are nonblack nonravens. To resolve Hempel's paradox, we must decide why this is so.

Stream of Consciousness

Try a different tack. Categories like "nonravens" and "nonblack things" are unnatural. Most of the time you are first aware that a "thing" is a raven or a herring or a steak knife. You don't naturally experience objects as "nonravens" or "nonherrings" or "non-steak knives." Only Hempel's original formulation ("All ravens are black") dovetails with how people really think.

Your train of thought *is* quite different with the two versions of the hypothesis. When you see a raven, your thoughts normally run like this:

(a) Look, there's a raven.
(b) And it's black.
(c) So it confirms the statement "All ravens are black."

Connecting a red herring to Hempel's hypothesis requires a more roundabout stream of consciousness!

(a) There's a herring.
(b) It's red.
(c) Oh, wait, how does that raven paradox go? Yeah, it's a "nonblack thing" . . .
(d) . . . and it's not a raven.
(e) So it confirms the statement "All nonblack things are nonravens" . . .
(f) . . . which is the same as "All ravens are black."

Between steps (a) and (b) in the original formulation—the instant after you realize that the object is a raven, but before you think about its color—the hypothesis is at risk. In that split second, the raven could be some other color and disprove the statement. The statement "All nonblack things are nonravens" is never really at risk in the second formulation. By the time you reach (c), you have already realized that the object is red (you deduced it was nonblack from the knowledge that it is red) and that it is a herring (you probably knew that all along).

Why is "raven" a reasonable category and "nonraven" not one? Well, ravens share many attributes in common, whereas "nonraven" is just a catchall term for anything that doesn't qualify. One category is figure and the other is ground. It's like the joke about a sculptor chiseling away everything that doesn't look like his subject. Sculptors don't think that way, and neither do scientists.

There is also a staggering numerical imbalance between the categories. Let's have one more go at the original idea: that the paradox has something to do with the relative numbers of ravens and nonblack things.

Infinitesimal Confirmation

Hempel's reasoning need *not* lead to a paradox when the number of objects under investigation is clearly finite. Suppose that all that existed in the universe was seven sealed boxes. Unknown to you, five of the boxes contain black ravens; one contains a white raven; and one contains a green crab apple. Then you could reasonably feel that opening a box and finding the crab apple confirms "All ravens are black." In fact, the speediest way to prove or refute the hypothesis would be to inspect all the nonblack things. There are only two nonblack things vs. six ravens. Of course, this model is artificial. It assumes prior knowledge of the number of things being investigated. You hardly ever know that, at least not at the start of the investigation.

More typical is the case where the original and not the contrapositive hypothesis talks of a knowably finite class of objects. The time, effort, and money required to establish "All ravens are black" is tied to the number of ravens (or the number of nonblack things). According to R. Todd Engstrom of Cornell's Laboratory of Ornithology, the world population of common ravens is something like half a million. More troublesome is the number of nonblack things. It is astronomical.

One day it is discovered that there is a monster in Loch Ness. There's just one monster; sonar equipment has established that there are no more of its kind. You want to test the hypothesis "All Loch Ness monsters are green." You approach the monster in a submarine, switch on the searchlights, and look out the porthole. The monster is green. Since there are no more Loch Ness monsters, the statement "All Loch Ness monsters are green" is thereby proven.

Here a single test of a hypothesis holds a lot of weight. There is only one chance for a nongreen monster to disprove the hypothesis. Taking the contrapositive seems even more ridiculous here than with the ravens. The contrapositive is "All nongreen things are non-Loch Ness monsters." Imagine going around and assigning a number to every nongreen thing in the world. Nongreen thing #42,990,276 is a blue lawn elf. Is it a non-Loch Ness monster? Yes! It supports the hypothesis. . . .

This is a woefully roundabout approach. Still supposing that there is just one Loch Ness monster and thus one potential counterexample, the chance that that arbitrary nongreen thing #42,990,276 is going to disprove the hypothesis is no greater than $1/N$, where N is the number of nongreen things. There might be something like 10^{80} atoms in the observable universe (which is the number written by putting 80 zeros after a "1"). There are at least that many nongreen objects. You might even claim that abstractions like numbers qualify as nongreen objects. Then the number is infinite.

This reasoning, anticipated by Hempel in his original musings in the 1940s, is very tempting. Possibly, a red herring *does* confirm "All ravens are black"—but only to an infinitesimal degree, because there are so many nonblack things. Checking the color of ravens is simply a more efficient way of confirming the hypothesis. In this vein, philosopher Nicholas Rescher estimated the costs of examining a statistically significant sample of ravens and nonblack objects. Rescher put the research tab at $10,100 for the ravens vs. $200 *quadrillion* for nonblack objects!

There remains the dilemma of how a red herring can confirm "All ravens are black" *and* "All ravens are white." You can try to picture it as being something like the mathematics of infinitesimals. The confirmation provided by a red herring for "All ravens are black" is on the order of $1/\text{infinity}$. The "infinity" in the denominator refers to the infinity of nonblack objects, of which the red herring is one. Since the herring is also a nonwhite object, it should confirm "All ravens are white" to an identical degree of $1/\text{infinity}$.

One divided by an infinite quantity is defined to be an infinitesimal, a number greater than zero but smaller than any regular fraction.

Does infinitesimal confirmation make the conflict any more palatable? We would be saying that a red herring confirms both "All ravens are black" and "All ravens are white," but only to an infinitesimal degree.

A small truth is still a truth; a small lie is yet a lie; and a contradiction is still a contradiction, even on an infinitesimal scale. The only out is to admit that the confirmation in both cases is precisely zero—as plain horse sense demands. Why, then, isn't a confirming instance of a hypothesis a confirming instance of its contrapositive?

The Paradox of the 99-Foot Man

Sometimes one paradox suggests the resolution of another. Paul Berent's paradox of the 99-foot man is another demonstration of the fallibility of Nicod's criterion. Say you subscribe to the reasonable belief: "All human beings are less than 100 feet tall." Everyone you've ever seen is a confirming instance of this hypothesis. Then one day you go to the circus and see a 99-foot-tall man. Surely you leave the circus *less* confident that all people are less than 100 feet tall. Why? The 99-foot man is yet another confirming instance.

There are two sources of this paradox. First, we don't always say what we mean. Sometimes the words we use imperfectly express the (often vague) hypothesis in our heads.

Chances are, you meant that no human being attains fantastic height; height an order of magnitude or more greater than the average. The precise figure of 100 feet was not vital. It was pulled out of the air as an example of the great height that you thought was definitely out of the question.

Had you been using the metric system, you might have said, "All human beings are less than 30 meters tall." Thirty meters comes to 98.43 feet, so the 99-foot man *would* be a counterexample to the 30-meter hypothesis. One feels that what you meant by saying "All human beings are less than 100 feet tall" is partially violated by the 99-foot man. It is like obeying the letter but not the intent of the law.

There is another root of the paradox. Have the hypothesis be the substance of a running bet you have with a friend. If ever a 100-foot-or-taller person turns up, you lose and owe your friend dinner at a fancy restaurant. The hypothesis is propounded not out of intellectual curiosity but solely to formalize the bet. Only the exact terms of the wager count. The 99-foot man is close but no cigar. He poses no threat whatsoever of deciding the wager against you.

You would still feel that the 99-foot man hurts the chances of your hypothesis being right. This is because you know many facts about human growth and variation that allow you to deduce an increased likelihood of a 100-foot person from the fact of the 99-foot man. Nearly every human attribute recurs (even to a greater degree) eventually. The 99-foot man demonstrates that it is genetically and physically possible for a person to attain a height of about 100 feet.

Now imagine that you find a way to test your hypothesis without acquiring any nonessential information. At the busiest part of Fifth Avenue, you place a sensor in the sidewalk that detects whenever anyone walks over it. A hundred feet above the sidewalk sensor is an electric eye. When someone steps on the sensor, the electric eye determines whether a beam of light 100 feet above the sidewalk has been broken by a tall pedestrian. A recording device keeps track of the total pedestrian traffic and the 100-foot-or-taller pedestrian traffic.

You check the meter to see the results. The readout is "0/310,628"—meaning that 310,628 pedestrians have passed, none (0) of whom was 100 feet tall. Each of the 310,628 pedestrians is a confirming instance of the hypothesis. Each confirms the hypothesis to a precisely equal degree. It would be ridiculous to say that some of the pedestrians provided more confirmation than others when all you know of the pedestrians is that they are shorter than 100 feet.

If it so happened that the 99-foot man crossed Fifth Avenue and was one of the people counted, he would confirm the hypothesis as much as anyone else, in your state of ignorance. Thanks to him, the meter reads "0/310,628" rather than "0/310,627," and you are slightly more confident for it.

Clearly, it is the additional information (that the man is 99-feet tall, and what you know about human variation) that converts a simple confirming instance into one that effectively disconfirms.

Philosopher Rudolf Carnap suggested that there is a "requirement of total evidence." In inductive reasoning, it is necessary that you use all available information. If you know nothing of the 99-foot man and only look at the meter readings, then he is a valid confirming instance. When you know more, he's not.

The requirement of total evidence has occasioned much soul-searching in the scientific community because it addresses much of the research arena of biochemistry, astronomy, physics, and other fields. The way we investigate genes or subatomic particles is more akin to the pedestrian traffic meter than simple observation. We do not meet RNA or quarks face to face; rather, we pose an exact question and learn the answers from machines.

Nothing is wrong with this, provided we do not limit our knowledge-gathering unnecessarily. If we are ignorant of other factors, and necessarily so, then we can generalize only from the information that is available. However, the more complete the information gathered, the more effective we are in making generalizations.

Ravens and Total Evidence

Let's recap. Science deals mostly in generalizations: "All X's are Y." Only through generalization can we compress our sensory experience into manageable form.

Generalizations are concealed negative hypotheses: "There is no such thing as a non-Y X"; or "The above rule has no exceptions." A generalization's contrapositive corresponds to the identical negative hypothesis.

In an infinite universe, proving a negative hypothesis is a supertask. (If the universe is merely finite but very big, proving a negative hypothesis is a herculean labor so close to a supertask as to make no difference.) We are incapable of supertasks, and have reason to be suspicious of knowledge attainable only through supertasks anyway.

Instead we establish generalizations through confirming instances: "X's that are Y"; black ravens in Hempel's example. This can never rigorously prove a generalization; only disprove it (through a counterexample: a nonblack raven). Tallying sightings of black ravens is a way of keeping score on how well established the hypothesis is. We feel that each black raven represents another instance in which the hypothesis was truly at risk of being disproved and came through unscathed. We do not feel that nonblack nonravens (confirming instances of the contrapositive) hold the same—or any—weight. And the puzzle of the ravens is to give a legitimate reason for this empirical instinct.

The requirement of total evidence is the key to unlocking the puzzle. Were our knowledge of the universe so poor that black ravens, nonblack ravens, black nonravens, and nonblack nonravens were nothing but data points, then it would be proper to confirm as the paradox suggests.

We know too much about ravens to confirm that way. Someone finds an albino crow (a counterpart of the 99-foot man). It's a nonblack thing and a nonraven. Far from confirming the "ravens are black" theory, it would cast strong doubt on it. Crows are in the same genus as ravens. If crows are prone to albinism, then possibly ravens are too. This background information negates the confirmation.

More generally, we know that ravens bear many, many more similarities to related birds than they do to red herrings or blue lawn elves. When this totality of evidence is taken into perspective, we realize it is a waste of time to examine nonblack nonravens. Whether all ravens are black is an issue best decided by observations of ravens and their relatives and by studies of biological variability.

Arguments from number of ravens vs. nonblack things are perhaps misleading. Consider again the case where the universe consists of seven sealed boxes, a case where most agree it is proper to count nonblack nonravens as confirming instances. Is the deciding difference between this and the real world truly one of number?

Picture a universe containing, say, 10^{80} sealed boxes. Most of the boxes contain black ravens; a few contain green crab apples; and maybe there is a white raven or two somewhere. You have opened a great number of boxes and thus far found only black ravens and green crab apples. Opening a new box and finding yet another black raven confirms "All ravens are black"—to a slight degree, for you have already opened a lot of boxes and there are trillions yet unopened.

Would not opening a box and finding a green crab apple slightly confirm the hypothesis as well? For one thing, it means one less possible refutation to worry about. For another, it increases your confidence that the objects you find in the boxes have certain fixed colors. You might even explain your faith in the hypothesis like this: "Every raven I've seen has been black. In fact, whenever I've seen something that wasn't black, it was always a crab apple, never a raven. The crab apples are 'the exception that proves the rule.' "

In this universe of sealed boxes, there is no ornithology, no albinism, no biological variation. In short, there is no background information about the way the world works. Instead of containing real ravens or crab apples, the boxes might as well contain slips of paper bearing the words "black raven," "white raven," etc. Now it is completely reduced to a formal game. If you open a box and find it to contain a slip of paper saying "white crow," there is no way of seeing that it has any different bearing on the hypothesis than "green crab apple."

We all know instinctively that it is wrong to ignore background evidence, but (before Hempel) this important fact went unrecognized in discussions of scientific method. It is unnecessary (logicians say impossible!) to deny the equivalence of the contrapositive. Hempel concluded simply that one must be wary of logical transformations of hypotheses. Yes, a contrapositive is equivalent, but confirmation does not always "recognize" logical transformations. The sundry ways in which consequences of inductive beliefs can mislead are the source of many paradoxes. A more troublesome paradox of induction is to come.

3

CATEGORIES

The Grue-Bleen Paradox

IN HIS ESSAY "The Analytical Language of John Wilkins," Jorge Luis Borges mentions a Chinese encyclopedia, the *Celestial Emporium of Benevolent Knowledge:* "On those remote pages it is written that animals are divided into (a) those that belong to the Emperor, (b) embalmed ones, (c) those that are trained, (d) suckling pigs, (e) mermaids, (f) fabulous ones, (g) stray dogs, (h) those that are included in this classification, (i) those that tremble as if they were mad, (j) innumerable ones, (k) those drawn with a very fine camel's hair brush, (l) others, (m) those that have just broken a flower vase, (n) those that resemble flies from a distance."

Man is the animal that invents categories. Science is a litany of phyla, genera, and species; eras and epochs; elements and compounds; leptons, mesons, and hadrons. The irony is that many of the categories judged scientifically valid appear just as arbitrary to the nonspecialist as the *Celestial Emporium*'s. Biologists divide the animal kingdom into about twenty-two phyla. Of these broad divisions, all "regular" animals (foxes, chickens, hippopotamuses, man) form a small subcategory of one phylum. Most of the other phyla enumerate the varieties of worms.

Borges's essay describes the ambitious and perhaps insane artificial language devised by British scientist and educator John Wilkins (1614–72). Wilkins's language divides the world into forty categories. Each category is divided into subcategories and sub-subcategories as in a library's cataloguing system. Wilkins associated a letter or two with each category. The words for things in Wilkins's language are assembled by stringing together the letters for the successive categories that define it. It is as if the title of each book in a library was also its catalogue number; or as if people's names were composed of letters from their ancestors' names.

"The word *salmon* does not tell us anything about the object it represents; *zana,* the corresponding word, defines (for the person versed in the forty categories and the classes of those categories) a scaly river fish with reddish flesh," Borges writes. "Theoretically, a language in which the name of each being would indicate all the details of its destiny, past and future, is not inconceivable."

Grue Emeralds

In 1953 American philosopher Nelson Goodman posed what he called the "new riddle of induction." The "grue-bleen" paradox, as it is better known, challenges our thinking about categories. A jeweler examines an emerald. "Aha," he says, "another green emerald. In all my years in this business, I must have seen thousands of emeralds, and every one has been green." We think the jeweler reasonable to hypothesize that all emeralds are green.

Next door is another jeweler having equally comprehensive experience with emeralds. He speaks only the Choctaw Indian language. Color distinctions are not as universal as might be thought. The Choctaw Indians made no distinction between green and blue—the same words applied to both. The Choctaws *did* make a linguistic distinction between *okchamali,* a vivid green or blue, and *okchakko,* a pale green or blue. The Choctaw-speaking jeweler says: All emeralds are *okchamali.* He maintains that all his years in the jewelry business confirm this hypothesis.

A third jeweler speaks Gruebleen, a strange, Esperanto-like language. The Gruebleen language has its own terms for colors, just as English and Choctaw do, but it has no word for green as such. Instead it has the word "grue." "Grue" can be defined in English as follows: If something is green before midnight, December 31, 1999, and blue thereafter, then it is grue. The Gruebleen-speaking jeweler naturally concludes that all emeralds are grue.

Put the question "What color will this emerald be in the year 2000?" to the jewelers. All three shake their heads and say they never knew an emerald to be any other color than what it is right now. The English speaker confidently predicts that the emerald will be green in the year 2000. The Choctaw speaker says it will be *okchamali.* The Gruebleen speaker asserts that the emerald will be grue in the year 2000 . . . but wait! "Grue in the year 2000" means blue in plain English. (It means *okchamali* in plain Choctaw.)

The paradox is that all three jewelers have had identical experiences with emeralds, and all have used the same inductive reasoning. Yet the Gruebleen speaker's prediction is at odds with the English speaker's. (The Choctaw speaker's prediction is compatible with either of his fellow jewelers'.) The paradox can't be swept away as meaningless. Come the turn of the century, at least one prediction will be wrong.[1]

The paradox can become as absurd as you wish. Let "grurple" mean green before a designated "zero hour" and purple thereafter. Let "emerow" mean something that is an emerald before the zero hour and a cow thereafter. Then the green emerald confirms "All emerows are grurple," which is to say, the green emerald will be a purple cow in 2000 A.D. By suitable choice of terms and zero hour, *anything* confirms that it will be *anything* else at *any* later time.

Gerrymander Categories

As with Hempel's paradox, there is an obvious "resolution" that fails miserably. The problem sure seems to be that gimmicky word "grue." "Grue" is inherently a more complicated word than "green"—look at its definition above! Grue is a "gerrymander" category, to borrow a term from politics. It has no natural significance;

[1] Gemological note: A blue emerald is paradoxical indeed, for emeralds are in fact transparent beryls that happen to be green from traces of chromium. A blue beryl of gem quality is called an aquamarine. Much rarer than true emeralds are "oriental emeralds," the green phase of corundum. (Rubies and sapphires are the red and blue forms.) With either the beryl or the corundum type, a nongreen emerald is self-refuting, like an orphan with parents.

it was constructed by Goodman with the sole aim of creating a paradox. It makes irrelevant reference to a specific point in time.

We do use some rather artificial categories in the real world. When a person in Chicago says it is 5 o'clock, he is actually saying that it is 5 o'clock in the region west of 82.5 degrees west longitude and east of 97.5 degrees west longitude, except as these boundaries are amended by local observances of Central Standard Time. It's 6 o'clock in the Eastern time zone, 4 o'clock in Mountain Time, and assorted other times at other places in the world. It's every hour—someplace—all the time. This definition sounds at least as cockeyed as that of "grue." It makes reference to geographic location, which is irrelevant to what time it is.

How much more sensible it would be to use Greenwich Mean Time all over the world. When it was 5:30 P.M. in São Paulo, it would also be 5:30 P.M. in Tokyo, Lagos, Winnipeg, and everywhere else. We might then regard the current method of stating time a patchwork out of a logic paradox.

And is "green" any less arbitrary? As logician W. V. O. Quine has pointed out, the concept of color is, to a physicist's way of looking at things, arbitrary. Light comes in a continuum of wavelengths, and there is no special distinction to those wavelengths that we call "green." Were we explaining what "green" means to a being from another planet, we would have to say something like "Green is what we experience when viewing light of wavelength greater than 4912 angstrom units but less than 5750 angstrom units." Why 4912 and 5750 rather than some other cutoff points? No reason—that's just the way things are.

Of course, "grue" inherits the spectral arbitrariness of "green" (and "blue"). "Grue," however, is arbitrary in a way that "green" is not. "Grue" supposes a *change* in color. It is not that nothing in the world changes from green to blue. Unripe blueberries do. But a simultaneous and universal change is quite unprecedented. "Grue" asks us to believe in this change, a change that has never been observed.

This sounds like a strong objection. But it sounds just as sensible inside out from the other side of the looking glass. The third jeweler's idiosyncratic tongue has another color word, "bleen." Something is bleen if is blue until midnight, December 31, 1999, and green thereafter.

In order to explain the English word "green" to the Gruebleen-speaking jeweler, we have to say that something is green if it is grue before midnight, December 31, 1999, and bleen thereafter. To him, raised on grue and bleen from the cradle, "green" is the artificial term. It is *green*'s definition that makes reference to a specific time.

The cross-definitions are as symmetric as bookends. Look in an English-Gruebleen/Gruebleen-English dictionary, and count the number of words in the definitions of "green" and "grue." "Grue" can be defined using "green" and "blue," or "green" can be defined using "grue" and "bleen." Asking which is the more fundamental term is like asking whether the chicken or the egg came first.

To get the full impact of this, imagine that "grue" and "bleen" are not made-up terms of a logic paradox but the very real terms of a natural language. Native speakers routinely say grass is grue and the sky is bleen. To them, saying that a dress is bleen does not raise the question of how, physically, it is going to turn blue at the turn of the century. (Any more than our saying that a banana is yellow implies that it will never turn brown but will be yellow forever.) When they say a dress is bleen it is because it is being perceived as bleen *right now*. It is the same color as the portion of a color chart that is labeled "bleen"; the color of the bleen sky or the first bleen-bird of spring. The only difference between their bleen and our blue is that this time clause is built into the definition (or is it?).

Counterfactuals

The grue-bleen paradox is partly about *counterfactuals:* terms that talk about what *would* happen even though it *hasn't*. A paper clip is flexible, acid-soluble, and meltable. It is all these things even if it is never bent, dissolved in acid, or melted. A grue emerald would be grue even if it is destroyed before 1999.

Counterfactuals abound in science. Following Goodman, astronomers might call the color of the sun "yelite." The sun is an average yellow star now, and will be a white dwarf in about 10 billion years. Of course, no one has ever observed the sun change from a yellow star to a white dwarf. No one has observed any star do that. All our direct experience confirms that the sun will be yellow forever as well as that the sun is "yelite."

What is the difference between this and Goodman's paradox? The astronomers' belief in the future change is not accidental. It is not the result of there happening to be a term "yelite" in somebody's dictionary. It is based on astrophysical theory that has been confirmed in other realms.

Terms like "grue" and "bleen" are suspect because they arbitrarily delay refutation until a future time. No possible experiment conducted in the twentieth century can distinguish a grue emerald from a green one. The future color change is a supposition that is (so far) unnecessary. For that reason we are justifiably suspicious of someone advancing a hypothesis that emeralds are grue.

True as this is, it does not make the paradox vanish. Again, the infernal symmetry of the situation rears its double head. The Gruebleen-speaking jeweler can complain that no twentieth-century experiment will aid him in deciding whether a grue emerald will turn bleen in the year 2000 (this being his definition of "green"). To resolve the paradox, even partially, you must find a way in which the situation *isn't* symmetrical.

The Rotating Color Wheel

Maybe the problem is the *suddenness* of the change. Sudden changes generally require a cause. In a vacuum, an object may continue its motion forever, but an abrupt change of velocity can come about only through an outside agent.

If the suddenness bothers you, let "grue" describe a gradual change from green to blue over a period of a thousand years. Better yet, suppose that all the colors are changing. The artist's color wheel is slowly rotating, so to speak, and what is green now will be blue in a thousand years, purple in 2000 years, red in 3000 years, and come full cycle back to green in 6000 years. "Grue" applies to that class of objects (emeralds, summer leaves, etc.) that are green now, blue in a thousand years, and so on.

Assuming a 6000-year cycle, the color of all things would change ever so slightly with each passing moment. The cumulative color change in a human lifespan would, however, be so minor that hardly anyone would notice. (The wizened gemologist who complained that emeralds aren't quite the color they were in his youth would be humored. Don't many old-timers think that winters are warmer now, baseball players inferior, etc., etc.?)

Nor could we infer the color shifting from historical evidence. Today's green emerald would have been yellow when it graced a feudal lord's ring and orange in Cleopatra's crown. But how do we know what the ancients meant by their color terms? If the classical authors used such and such a word for the color of emeralds and grass and the Atlantic Ocean, we would translate it as "green." It *might* be that if we took a time machine back to the year 1, we would find all these things orange to our eyes. We cannot be certain that the Old English "grene" wasn't actually yellow either.

The Inverted Spectrum

This idea blends with "inverted spectrum" thought experiments, much discussed by philosophers. Suppose that you have, from birth, seen colors exactly opposite from everyone else. That is, the color sensation you get when you look at a Red Delicious apple—the sensation you have been trained to call "red"—is actually what everyone else calls "green." All the colors you see are the opposite of everyone else's. Is there any way for two people to describe their subjective color sensations to each other in such a way that they can be *sure* they see the same colors?

It seems to be impossible. Colors are usually described by comparing them to something else (turquoise blue, brick red, ivory, etc.). This would not work, for the same reason mentioned above. The best case for believing that we might be able to detect inverted color sensations is the supposed correlation of colors with psychological states. We are told that light blue and light green are restful; that red incites anger or hotheadedness; that blue is for boys and pink is for girls; that some colors (blue maybe) are more popular or tasteful than others (orange and purple maybe).

It is possible that certain colors have intrinsic psychological effects that have developed through evolution. On the other hand, it is possible that these are only societal conventions that children internalize at an early age. Unlike a lot of these epistemological issues, the inverted-spectrum debate might be settled by starting a country somewhere with all the colors inverted (by extensive use of nontoxic dyes?!). Green vegetables would be dyed bright red (but still called "green vegetables"); baby clothing would be "blue" (really orange) for boys, "pink" (really olive) for girls, and so on. Companies importing artists' pigments would have to squeeze violet paint out of its original tubes and put it in tubes labeled "yellow." Color photographs from the outside world would be permitted, but only the negatives! The "country" would be self-contained and underground, so that the blue of the sky wouldn't contaminate the experiment. Would people raised in this country share our color preferences? Would an indigenous work of abstract art be detectable in any way?

Whether or not it is completely impossible to detect an inverted spectrum, it is certainly difficult. So we can have a gradual grue-bleen paradox in which there is no sudden change and no unobserved future change. The alleged "change" is and has been going on all the time, and all our present and historical experience is compatible with the change. That seems to rule out any easy resolution of the paradox.

The inverted spectrum and the grue-bleen paradox probe a lot more than colors. Goodman used colors as an example of the categories into which we divide the world. Through categories, experience melds with language. Goodman's jewelers hold empirical beliefs about emeralds that have stood the test of time—and the beliefs are radically different!

Demon Theory No. 16

Instinctively, we know that "All emeralds are green" is a good hypothesis, and "All emeralds are grue" is somehow flawed. The question is how we distinguish reasonable hypotheses from unreasonable ones. You might answer, "By doing experiments, of course!" That is one way of distinguishing, but scientists cannot test every hypothesis, good, bad, or indifferent.

"Research is the process of going up alleys to see if they are blind," joked biologist Marston Bates. The role of scattershot research is severely limited, however. Philosopher of science Hilary Putnam illustrated this point with his "demon theory." The theory (really a hypothesis) is this: A demon (maybe Descartes's) will appear before your eyes if you put a flour bag on your head and rap a table 16 times in quick succession. Now, of course, this is stupid, but it *is* a hypothesis and *is* capable of being tested. It can be tested a lot easier than most scientific hypotheses can be tested.

The above is Demon Theory No. 16. There is a Demon Theory No. 17, which is the same except that there have to be 17 raps, and a Demon Theory No. 18, and a Demon Theory No. 19, etc. There is an infinity of demon theories. Obviously, said Putnam, scientists have to be selective about the theories they test. You could spend your life testing dumb theories and never get anywhere. It is vital that you winnow out the "possibly true" hypotheses from the "not worth bothering with" ones before getting as far as experimentation.

Unlike Putnam's demon theory, most hypotheses are motivated by experience. A snowflake falls on your sleeve. It has six sides. A reasonable hypothesis is "All snowflakes have six sides." But why that and not "All snowflakes that fall on Tuesday have six sides" or "All things have six sides" or "Everything that melts has an even number of sides" or "All hexagonal objects have six sides?" More important, why do we even think that something about the shape of snowflakes is generalizable? The very fact that there is a word "snowflake" presupposes the common knowledge that there is a class of tiny, cold, white objects that fall from the sky and that may have other properties in common. Without the implicit hint given by the word, one might grope for hypotheses such as "Everything in the class consisting of (this white thing on my sleeve, Queen Victoria, lasagne, and all the beach balls in the Southern Hemisphere) has six sides."

Anything Confirms Anything

Bad hypotheses have a way of subverting evidence. An example is the paradox often called "anything confirms anything"—a paradox that probably occurs, in the form of fallacious reasoning, more frequently than any other of those discussed in this book.

It is reasonable to think that something that confirms a hypothesis will confirm any necessary consequence of that hypothesis. If man is descended from apes, then undeniably Darwin is descended from apes. A fossil that confirms the hypothesis that man is descended from apes must also confirm that Darwin is descended from apes. So far so good.

Take a compound statement such as "8497 is a prime number and the other side of the moon is flat and Elizabeth the First was crowned on a Tuesday" (this example from Goodman). To test this, you check 8497 for divisors and conclude that is it prime. This discovery confirms the compound statement, and one consequence of the compound statement is that the far side of the moon is flat. The fact that 8497 is prime confirms that the moon is flat!

Of course, the compound statement could lump together any propositions at all. Replace them with propositions of your own choice and make your own paradox. Anything can be shown to confirm anything else.

Evidently it is easier to join hypotheses logically than to be sure that there is valid reason for linking them. This link is essential for valid confirmation. Goodman's sentence is patently a hodgepodge, but it suggests the wide-ranging consequences of any powerful theory. Many proponents of pseudoscience use the "anything confirms anything" argument. To give just one popular example:

HYPOTHESIS: Clairvoyance exists *and* it's possible because there's a lot that physicists don't know about cause and effect.

EVIDENCE: Bell inequality experiments, which seem to show instantaneous communication between subatomic particles.
CONCLUSION: Bell inequality experiments confirm the hypothesis, so they support the existence of clairvoyance!

Ockam's Razor

There is an aesthetic to science. The "beauty" of a theory is measured largely by its simplicity. A simple theory that explains a lot is preferred to a complicated theory that explains little—even though, on the face of it, there may be no particular reason to believe that the complicated theory is any less right than the simple one.

This important principle is called "Ockam's razor." The name comes from William of Ockam (the name is also spelled Occam and Ockham), a Franciscan monk born about 1285. (Very similar doctrines were propounded earlier by Duns Scotus and Odo Rigaldus.) A controversial figure embroiled in disputes with popes and antipopes, Ockam was one of the most influential of medieval thinkers. He died, probably of the plague, in 1349.

Ockam is best known for something he may never have said: *Entia non sunt multiplicanda sine necessitate,* or "Entities are not to be multiplied beyond necessity." The sentiment, if not those words, is his. He meant that you should not resort to new assumptions or hypotheses (entities) except when necessary. If a footprint in the snow *might* be explained by a bear, and *might* be explained by a previously undiscovered manlike creature, the bear hypothesis is favored.

The principle can be misunderstood. It is not a matter of choosing the less sensational explanation. One favors bears over abominable snowmen only when the evidence (such as a half-melted footprint) is so deficient that both the bear and the yeti theory account for it equally well.

Ockam's razor is fallible. It has often favored a *wrong* hypothesis. Is the earth round? Do tiny living creatures cause disease? We now know that these hypotheses account for observations very well, but at some point the Ockam's razor principle rejected them. A notorious case of misplaced skepticism (often cited by proponents of ghosts, UFOs, and other currently unaccepted beliefs) is the French Academy's prolonged rejection of the reality of meteorites. On the finest scientific advice, dozens of meteorites in European museums were thrown out as superstitious relics.

Here we come to one of the most troublesome points of confirmation theory. In every scientific discovery, there is a stage where two competing theories account for observations about equally well. There is often a simpler hypothesis, A, which everyone has been believing all along, and a new hypothesis, B, which postulates some new "entity," in Ockam's words. Theory A could be the belief that the earth is the center of the universe, and B could be Copernicus's heliocentric theory. Or to take an example less obviously stacked in favor of B, A could be that there are no UFOs, and B could be that UFOs exist. When does the evidence justify the new entity?

It is difficult to give a hard-and-fast answer, for we all believe many things on the basis of slight evidence. If you glance at the cover of a tabloid in a supermarket and read that a prominent actress has eloped, you probably take it for a fact. If the same tabloid the next week says in equal-sized print that UFOs abducted a woman in Arizona, you probably don't believe it. As astronomer Carl Sagan points out, there is a significant yet usually unconscious rule of confirmation at work here: The more outrageous the hypothesis, the more evidence is needed to confirm it.

The rationale is that a prosaic hypothesis is partially confirmed by all our prior knowledge of similar occurrences. An incredible hypothesis is not. This, however, raises the possibility of being tricked into believing a series of wrong prosaic hypotheses over a less prosaic truth (as in the French Academy's dismissal of meteorites). There is, for instance, a lot of evidence for the existence of ghosts. Many thousands of people have reported seeing them, and not all of them are kooks; there are even some fuzzy photographs. There is no categorical explanation for the reports of ghosts (other than that ghosts exist). It is maintained that there is always a "logical explanation," but this explanation is in one case a branch scratching against a window, in another a hallucination, in another mice in the attic, in still another a hoax. In yet other cases, none of these explanations can be offered, but still it is maintained that some cause not involving the paranormal exists.

In sheer quantity, the evidence for ghosts is probably greater than that for the existence of will-o'-the-wisps, the strange lights seen over marshes. Yet science believes in will-o'-the-wisps and not in ghosts. Ultimately, more theories are refuted by the poor quality of their own evidence than by contrary evidence. There is usually something wrong with a theory that has lots of supporting "evidence," all of it dubious. This seems to be the case with the theory that there are ghosts. On the other hand, some of the time will-o'-the-wisps are visible for all to see.

But in Goodman's paradox, we are skeptical of one hypothesis (emeralds are grue) even though it has precisely the same supporting evidence as another (emeralds are green). The problem rests with the hypothesis, not the evidence.

"All emeralds are grue" speaks of an entity, grueness, that we can do without. Invoking Ockam's razor, we can say, "Hold it! We already have all the color words we need. It is pointless to add a term like 'grue' until you produce something that is grue (and not just green)."

BUT—once again—the Gruebleen speaker can throw these words right back in our faces. He has all the colors he needs, has no need for "green," won't have until he sees something that actually is green (not grue).

Lively debate on the grue-bleen paradox continues. For the time being, most analyses concur that our preference for "green" rather than "grue" is based on simplicity. The difficulty is finding a way out of this vicious cycle whereby the Gruebleen speaker can parrot our every argument! Here is one way.

The Day of Judgment

Ask yourself what will happen on the semantic day of judgment, January 1, 2000 A.D. There are four possibilities.

1. Everyone might wake up and find that the sky is green and the grass is blue! We'll realize that "green" was the misleading term and "grue" was right—

Otherwise, the speakers of Gruebleen will have to accept the sameness of colors after the zero hour in one of three ways:

2. Gruebleen speakers could wake up and be *surprised* to find that the (still blue) sky has "changed" from bleen to grue. This is what Goodman facetiously implied.

3. Or Gruebleen speakers could go to bed the previous night *fully expecting* the "change." It would be like resetting a watch for daylight saving time or travel across time zones. The Gruebleen speakers would realize that their color terms don't jibe with the way the world works.

4. Finally, the Gruebleen speakers might not recognize the "change" at all (through failure to understand the time clause in the definitions of "grue" and "bleen"). For *how do Gruebleen-speaking parents teach their children the language?*

Many philosophers believe that no one could really learn the Gruebleen language as their first language. Sure, parents would point to the grass and say "grue," point to the sky and say "bleen." But there is more to grue and bleen than that. The perceptual change (you shouldn't call it a color change, since grue and bleen are colors to those using those words) at midnight, December 31, 1999, has to be communicated at some point in the process of acquiring language. At some point, a parent or teacher has to sit a Gruebleen-speaking child down and tell him the facts of grue and bleen.

Here the symmetry breaks down. No one has to tell an English-speaking child that green things *don't* turn blue in the year 2000 to prevent him from getting a wrong idea of what "green" means. It comes naturally. There is an irrelevant reference to time in the definition of "grue" after all.

Projectability

Goodman's riddle has radically changed thinking on induction. Goodman talked of the "entrenchment" of terms in language. There is a word for green and not for grue because one agrees with the way of the world and the other doesn't. The differing color distinctions of some other natural languages agree with Goodman's precept. Choctaw does not make our distinction between green and blue, but no single word of any natural language means grue or anything like it.[2]

A problematic attribute like grueness is said to be *nonprojectable*. An attribute is projectable if it can be used validly in inductive reasoning. Greenness is projectable, in that an instance of a green emerald confirms the obvious generalization that "all emeralds are green."

There are, however, three types of situations where positive instances of a hypothesis are nonprojectable. One is that of the grue-bleen paradox; another is "anything confirms anything."

The third nonprojectable case is a lemma to the grue-bleen paradox. Consider the hypothesis "All emeralds have been observed." Every emerald ever seen has been observed, of course. Projecting all these instances of observed emeralds leads to the absurd conclusion that we have observed all the emeralds that exist—that there are no

---

[2] Because of the wide discussion of Goodman's paradox, "grue" and "bleen" have entered the English language and are likely to appear in future unabridged dictionaries!

unseen emeralds. In this case, there is nothing artificial about the attribute of being observed. "Observed" is entrenched in the language as well as "green."

Are Quark Colors Grue-ish?

Scientists must be wary of nonprojectable terms. Quarks are hypothetical entities said to reside deep inside protons, neutrons, and other subatomic particles. Quarks are counterfactual: Not only has an isolated quark never been observed, but (under most theories) an isolated quark is impossible. Quarks are what a proton *would* split into, *if* it could be split, which it *can't*.

Quarks are held inside protons and neutrons by a "color force." Most physical forces, like gravity and electric attraction, decrease with distance. The color force does not decrease with distance. It is as if all quarks are connected with rubber bands that continue to exert force from any distance. Consequently, it would take an infinite expenditure of energy to free a quark permanently from a proton. Even a less ambitious project, like pulling a quark an inch out of a proton, would take a fantastic amount of energy (and wouldn't succeed in any case—the energy would create new particles rather than a bent-out-of-shape proton).

The way nature seemingly conspires to avoid free quarks has always been suspicious. Some fear that unseen quarks may be something like the unseen blue of twenty-first-century grue emeralds. Although the theory of quarks and the color force—quantum chromodynamics—has been confirmed in many ways that the grueness of emeralds has not, controversy rages over whether quarks are "real" particles or only a convenient shorthand for categorizing the particles said to be made up of them.

Identical questions were raised about the reality of atoms in the nineteenth century. But John Dalton's atomic theory did not rule out detection of individual atoms—and such confirmation came eventually in the gold-foil experiments of Ernest Rutherford (1911).

Moreover, there is a weary exasperation with the increasingly complex quark model. There are different varieties of quarks. These varieties were called "colors" and "flavors." (They have nothing to do with real colors, much less flavors, but what can you call attributes of something so removed from the world of the senses?) There are three colors (called red, blue, and green) and six flavors (called up, down, strangeness, charm, bottom, and top). That creates eighteen types of quarks, not counting antiparticles. Then there are electrons, neutrons, gluons, Higgs particles . . .

Some wonder if colors and flavors may be artificial complications of a simple reality we do not yet understand. Possibly someday someone will hit on how things really are, and we will realize that our current physics is a strained way of describing this reality. We could be like a Gruebleen speaker trying to understand why the sky has turned grue on judgment day. The answer is not in the sky, but in our heads.

4

THE UNKNOWABLE

Nocturnal Doubling

SUPPOSE THAT LAST NIGHT, while everyone slept, everything in the universe doubled in size. Would there be any way of telling what had happened? So runs one of the most famous intellectual riddles of all time, posed by Jules Henri Poincaré (1854–1912), a talented popularizer as well as an eminent scientist of his day.

One's first impulse is to assume that any drastic change like that would readily be detectable. Think again: Since *everything* has doubled in size, so have all rulers, yardsticks, and tape measures. You cannot *measure* anything and detect the change.

The vaunted platinum/iridium bar in a suburban Paris cellar, the original basis of the metric system, has doubled too and provides no clue to the change. The meter is currently defined as the length of 1,656,763.83 wavelengths of a specific orange light given off by krypton gas. Still no good. The special fluorescent tubes containing this gas are twice as big, and so are the krypton atoms they contain. The electron orbits of the krypton atoms are twice as big, and therefore the light produced has twice the wavelength.

Wouldn't things *look* bigger? The picture on your bedroom wall is now twice as big. But your head is twice as far from the picture (from any particular point in the enlarged room). The two factors precisely eliminate any perceptual change.

All right, try this. You're in foggy London, looking at Big Ben. The clock is twice as big, and you're twice as far from it, from any given vantage point. The perspective is identical. But your line of sight traverses twice as much fog. Shouldn't Big Ben look hazier?

The trouble is, it's really the number of fog droplets that cause haze, and this number hasn't changed. The droplets are twice as big, and they scatter the doubled photons exactly the way they did before. Big Ben would look as clear or as hazy as it would have had there been no doubling. Similar arguments show that everything would look the same.

The real point of the thought experiment is this: Granted that it is impossible to detect the change, *is* there then a change at all? The question recalls the old metaphysical riddle asking if a tree falling in the forest with no one to hear it makes a sound.

You might say that the nocturnal doubling is real, in that God or some such being "outside" the universe would know of the change. You can imagine God sitting in hyperspace somewhere and watching our universe double in size. This misses the point entirely. Everything that exists must double in size, including God. Not even God can do anything to demonstrate the change. *Then* is the change real?

Antirealism

Poincaré said no. It is pointless even to talk of such a change, he felt. Here words are deceptive: "what if everything in the universe doubled in size" sounds like it describes a change, but the "change" is an illusion.

Others differ. Nocturnal doubling illustrates two competing schools of philosophy. The school called realism allows that nocturnal doubling might be real, even if unobservable. Realism holds that the external world exists independent of human knowledge and observation of it. There are truths beyond our recognition. These include not only truths that are currently unknown and seem impossible to discern (such as what happened to Ambrose Bierce or whether there is life on Alpha Centauri) but truths that no one will ever know, no matter what. The realist school says these truths still exist. Common sense is primarily realist: Of course the tree makes a sound, even if no one hears it.

Philosophers of the antirealist school argue that there are no evidence-transcendent truths (truths that cannot be demonstrated empirically). Given that no one could ever detect nocturnal doubling, it is absurd and misleading to say that the doubling occurred. Saying that everything doubled last night and saying that everything is the same size are (at most) just different ways of describing the same state of affairs.

A big part of philosophy is deciding which questions about the world are meaningful. Antirealism is the dogma that only those questions that may be settled on the basis of observation or experiment have meaning. It resists assuming anything about the unobserved and unobservable. Antirealism sees the world as something like a movie set where the buildings are just façades. It resists the temptation to fill out the buildings behind the façades.

The difference between what is unknown and unknowable may be subtle. No one knows Charles Dickens's blood type. The ABO blood types were not discovered until a generation after Dickens's death (by Austrian biologist Karl Lansteiner in 1900), and thus Dickens's blood type was never determined. Though the Dickens's blood type may remain forever unknown, most would feel that that doesn't change the fact that Dickens had *some* blood type.

In contrast, everyone recognizes as meaningless a question like "What was David Copperfield's blood type?" It is meaningless because the fictional character exists only insofar as Dickens imagined him, and Dickens did not give this information in the story. It is not merely that we are ignorant of Copperfield's blood type. There is nothing to be ignorant of.

Antirealism speaks of questions that are undecidable in principle, like nocturnal doubling. In its most radical form, antirealism is the belief that the unknowables of the external world are just as meaningless as questions about a fictional character's blood type. There is nothing to be ignorant of.

Were that all there was to it, the question of realism vs. antirealism would be purely a matter of philosophical preference. Actually, there are many open questions of physics, cognitive science, and other fields where the relationship between what is unknowable and what is meaningful blurs. This chapter will examine several of these variations on the unheard tree.

Physics Goes Haywire

There is more to the nocturnal-doubling debate. For one thing, not everyone agrees that a nighttime doubling would be undetectable. One of the best cases for detectability has been put forward by philosophers Brian Ellis and George Schlesinger.

In 1962 and 1964 papers, Ellis and Schlesinger claimed that the doubling would have a large number of physically measurable effects. Their conclusions depend on how you interpret the thought experiment, but they are worth considering.

For instance, Schlesinger claimed that gravity would be only one-fourth as strong because the earth's radius would have doubled while its mass remained the same. Newtonian theory says that gravitational force is proportional to the *square* of the distance between objects (in this case the center of the earth and falling objects at its surface). Doubling the radius without increasing the mass causes a fourfold reduction of gravity.

Some of the more direct ways of measuring this change in gravitational pull would fail. It wouldn't do to measure the weights of objects in a balance. The balance can only compare the lessened gravitational pull on objects against the equally lessened pull on standard pound or kilogram weights. Schlesinger argued, however, that the weakened gravity could be measured by the height of the mercury column in old-fashioned barometers. The height of the mercury depends on three factors: the air pressure, the density of mercury, and the strength of gravity. Under normal circumstances, only air pressure changes very much.

The air pressure would be eight times less after the doubling, for all volumes would be increased 2^3 times, or eightfold. (You wouldn't get the bends, though, because your blood pressure would also be eight times less.) The density of mercury would also be eight times less; these two effects would cancel out, leaving the weakened gravity as the measurable change. Since gravity is four times weaker, the mercury should rise four times higher—which will be measured as *two* times higher with the doubled yardsticks. That, then, is a measurable difference.

Schlesinger applied the doubling to some other standard physical laws and further claimed:

· The length of the day, as measured by a pendulum clock, would be 1.414 (the square root of 2) times longer.

· The speed of light would increase by the same factor, as measured by a pendulum clock.

· The year would contain 258 days (365 divided by the square root of 2).

You can quibble with some of Schlesinger's work. Schlesinger uses a pendulum clock as the unit of time. This clock is much slower because gravity is weaker and the length of the pendulum is doubled. Other types of clocks would not share this slowing. You can argue on the basis of Hooke's law (which governs the resistance of springs) that an ordinary watch with a mainspring would run at exactly the same rate after the doubling.

There is also the question of whether the usual conservation laws apply during the expansion. Schlesinger supposes that the angular momentum of the earth must remain constant (as it normally would in any possible interaction), even during the doubling. If the earth's angular momentum is to remain constant, then the rotation of the earth must slow.

There would be other consequences of conservation laws. The universe is mostly hydrogen, which is an electron orbiting a proton. There is electrical attraction between the two particles. To double the size of all atoms is to move all the electrons "uphill," twice as far away from the protons. This would require a stupendous output of energy. If the law of conservation of mass-energy holds during the doubling, this energy has to come from somewhere. Most plausibly, it would come from a universal lowering of the temperature. Everything would get colder, which would be another consequence of the doubling.

The thrust of Schlesinger's argument is this: Suppose we got up one morning and found that every mercury barometer in the world had shattered. Investigating further, we find that mercury now rises to about 60 inches rather than 30. (The barometers shattered because no one made the glass columns that long.) Pendulum clocks and spring-wound clocks now keep different time. The velocity of light, when measured by a pendulum clock, is 41.4 percent greater. The length of the year has changed. There are thousands of changes; it is as if all the laws of physics went haywire.

Then someone suggests that what's happened is that all lengths have doubled. This hypothesis accounts for all the observed changes, and makes many predictions about other changes. Upon hearing of the nocturnal-doubling hypothesis, specialists in the most recondite subfields of physics can say, "Now wait a minute, so-and-so's law, which talks about distances, would cause such and such to happen if it was true that all lengths have doubled." Every time such a consequence is checked, it is found to be accurate. The doubling theory would quickly be confirmed and established as scientific fact. Not only that—it would bid fair to be a paradigm of confirmation. It is hard to imagine any theory that would have so many independently checkable consequences.

Now for the nub of the issue. Since there is a conceivable state of affairs in which we would be forced to conclude that all lengths have doubled, and since that state of affairs does not currently obtain, we are correct in saying that everything *didn't* double last night.

Demons and Doubling

Schlesinger's point is well taken. It delimits rather than demolishes the intent of the original thought experiment. There are really two conceivable versions of Poincaré's thought experiment. You may find it helpful to think about it this way:

Imagine that the laws of physics are implemented by a demon who runs around the universe making sure that everything squares with said laws. The demon works like a cop on the beat, going from place to place and making sure that laws are observed.

The instant after the doubling, the demon is making a routine check of Newton's law of universal gravitation. This law says that the force *(F)* between any two objects equals the gravitational constant *(G)* times the product of their masses *(m_1* and *m_2)*, divided by the distance between them *(r)* squared:

$$F = \frac{Gm_1m_2}{r^2}$$

The demon is making sure that the earth and the moon obey this law. He measures the mass of the earth, the mass of the moon, and the distance between them. He looks up the value of the gravitational constant in his handbook. He punches these numbers into his calculator and gets the correct value for the gravitational force between the earth and the moon. Then he turns a dial on a control panel and sets the momentary strength of the gravitational force between the earth and the moon.

The question is: How does the demon measure distances? Does he just "know" them, and thus is magically aware of the doubling? Or is he in the same boat with us, and has to measure distances by comparing them with other distances?

If the demon knows about the doubling ("if the laws of physics recognize the doubling"), then we have Schlesinger's version of the thought experiment. That kind of doubling would be detectable, and since we have not detected it, we are justified in saying that a nocturnal doubling *visible to the laws of physics* has not occurred. If, on the other hand, the expansion is invisible even to natural laws, then there is no way of detecting the expansion. I think there is no question but that Poincaré would say he meant a doubling that was invisible to the laws of physics.

For the record, universal changes are not the sole province of philosophers. Physicist Robert Dicke has proposed a theory of gravity in which the gravitational constant changes slowly with time. It is clear from Poincaré's example that any useful hypothesis must have measurable consequences. Dicke's theory does. The gravitational constant measures the universal strength of gravity. If it doubled one night, you'd know it. The bathroom scales would tell you you weighed twice as much the next morning. Birds would have trouble flying; yo-yo's wouldn't work; in a myriad ways, the world would be a very different place. In fact, it's doubtful anyone could survive a doubling of the gravitational constant. The intensified gravity would compress the earth in a series of unfathomable earthquakes and tidal waves. The sun would shrink too, and burn hotter, searing the earth.

Dicke's theory suggests that the gravitational constant is decreasing, not increasing. A halving of the gravitational constant would have the opposite effects but probably would be as deadly: you'd weigh less; birds would soar higher than ever before; the sun would cool off as it bloated, and we'd freeze to death. Of course, in Dicke's theory the decrease is imperceptibly slow: maybe as much as a 1 percent decrease in a billion years. Even that slight change might be detectable in highly precise measurements of planetary motion and perhaps in geophysical effects. Attempts to find evidence of a change in the gravitational constant have so far failed.

Variations

Poincaré's thought experiment, which helped pave the way for the acceptance of Einstein's relativity, illustrates one of the most palatable forms of antirealism. Many variations on Poincaré's fantasy are possible. Obviously, a nocturnal shrinking of the universe would be equally undetectable. Would there be any way of telling if:

• The universe "stretched" to twice its size in one direction only (and things that changed their orientation after the change stretched or compressed proportionately)?
• The universe turned upside down?
• The universe became a mirror image of itself?
• Everything in the universe doubled in value, including money, precious metals, and whatever is used for currency on other planets?
• Time started running twice as fast?
• Time started running twice as slow?
• Time started running a trillion times slower?
• Time stopped completely, starting right . . . NOW!
• Time started running backward?

Most would say that all these scenarios would be equally undetectable and meaningless. The last two questions in the list deserve comment.

You could never know it if time stopped. You *can* know that it didn't stop last night or three seconds ago when you read the word "NOW!" (I speak of time stopping "forever," and not just stopping for a "while" and then starting back up again. A temporary suspension of time would be undetectable.) Whether this very moment is the moment in which time stops is something you cannot know until after the fact.

If you are sure that time is *not* running backward, ask yourself how you know this. You probably cite your memories of the past. It is now 1988. You have memories of experiences in 1987, 1986, 1985, etc.; you do not have memories of 1989, 1990, etc. But you would, for the moment, have the memories you do whether time was going forward or backward from 1988. The question is whether the moving finger of time adds to or deletes from this stock of memories. There is no way of telling!

Did Time Begin Five Minutes Ago?

The best-known thought experiment about time was devised by Bertrand Russell in 1921 (or was it?). Suppose that the world was created five minutes ago. All memories and other traces of "prior" events were likewise created five minutes ago as a private joke on the Creator's part. How do you prove otherwise? You can't, Russell insisted.

Few would argue with Russell, for the same reasoning quashes any proposed refutation. A 1945 bottle of Château Latour—the yellowed pages of a Gutenberg Bible, dated 1457—fossils—carbon 14 dating—astrophysical evidence of the age of stars—the Hubble time —none means anything more than the time on a clock in a painting. They were simply created that way, five minutes ago.

(In a weird case of theology imitating philosophy, some opponents of Darwinian evolution have claimed that God created fossils to tempt the wicked into disbelieving that the world was created in 4004 B.C., as Archbishop Ussher figured in the margins of the King James Bible.)

Perils of Antirealism

The antirealist position can be applied unwisely. There is always the danger of assuming too readily that something presently unknowable must remain so forever. In 1835 French philosopher and mathematician Auguste Comte, founder of logical positivism, advanced the chemical composition of stars as something necessarily beyond human knowledge. "The field of positive philosophy lies wholly within the limits of our solar system," he wrote.

Not only was Comte wrong; he was untimely. As he wrote those words, physicists were puzzling over the mysterious dark lines Joseph von Fraunhofer had discovered in the sun's spectrum. A generation later, Gustav Kirchhoff and Robert Bunsen realized that the lines were produced by chemical elements in the sun. When the spectroscope was pointed at a distant star, it would reveal the star's chemical composition as well.

One scientific topic often cited in discussions of antirealism is black holes. It is sometimes asserted that predictions about the interiors of black holes are unverifiable and thus open to the antirealist claim that the predictions are meaningless. This is not strictly true, and it may be useful to examine why.

A black hole is a region of space with such an intense gravitational field that nothing which enters can ever leave. The "nothing" is absolute and final. Neither matter nor energy of any kind can get out. Since even information must be conveyed in matter or energy, not even information can exit a black hole.

Think about that: There is no way for someone inside a black hole to send a radio signal to us; no way to put a message in a bottle and propel it out. We on the outside can never know anything directly about whatever goes on inside a black hole. Does it make any sense, then, to talk of what happens inside a black hole?

Black holes are a prediction of Einstein's theory of gravity, the general theory of relativity. This theory *does* make predictions about the interior of black holes—as well as virtually guaranteeing that those predictions can never be tested. A black hole results whenever a great enough mass is concentrated in a small enough space. When a large star (about twice the size of our sun or bigger) runs out of thermonuclear fuel and collapses, its own gravity will compress it ever smaller and smaller. The more compact it gets, the more intense its gravitational field becomes. Once the gravity passes a critical point, no force known to physics can stop it. The atoms are crushed out of existence. The star shrinks to (as far as anyone knows) a point.

Though the star disappears, its gravity remains. It leaves behind an intense gravitational field, the black hole. The "boundary" of a black hole is called the *event horizon*. This is the literal point of no return. Anything that plunges within this spherical boundary can never come out again.

A black hole would be spherical and typically only miles in circumference; it would be utterly, totally black; and it would warp the light of any objects behind it, something like a bubble in a pane of glass. A typical stellar black hole resulting from the complete collapse of a star twice the sun's mass would have an effective diameter of twelve kilometers (seven miles). This effective diameter is a fiction. To measure a black hole's diameter (or radius) you would have to stretch a measuring tape or equivalent *inside the black hole.* Any observer who did that could never report the measurement to the outside world. Besides, this diameter is theoretically infinite through the warping of space. What one can do is measure the circumference of a black hole. In principle you can do this by looping a measuring tape around the black hole, just outside the event horizon. Dividing this circumference by pi gives the effective diameter—a measure of the space the black hole seems to occupy, to observers in the outside world.

Black-Hole Probes

Let's examine several schemes for getting information out of a black hole. It would do no good to send in a NASA-style probe that relays data back via radio. Radio waves, like visible light, are a type of electromagnetic radiation. A radio signal can no more emerge from a black hole than can a flashlight beam.

Another easily dismissed scheme is to send a rocket in and out. Any planet or star has an escape velocity. A rocket must exceed the escape velocity to leave the body without falling back. For a black hole, the escape velocity is the speed of light. Since the speed of light is a cosmic speed limit that nothing can exceed, no possible rocket design could extricate itself from a black hole.

You can imagine rigging up a bathyscaphlike probe. The probe, fitted with searchlights and cameras, is lowered into a black hole on an absolutely unbreakable cable. The cable is anchored to—well, something really big and solid. The probe snaps some pictures, then it's pulled back out.

It wouldn't work. Once an atom of cable is inside the event horizon, no physical force, including the electromagnetic forces that hold matter together, can transport the atom outside again. There can be no such thing as an "absolutely unbreakable" cable in a universe that contains black holes.

Agreed, then, that nothing that goes into a black hole ever comes out again. That does not necessarily mean that predictions about black-hole interiors are unverifiable. In principle, a person could go inside a black hole and take a look around. He could never come out again, and he would not survive long on the inside. It would furthermore have to be a very large black hole, or the observer could not even survive crossing the event horizon.

The distortion of space around a black hole takes the form of an immense tidal force. This is the same type of force that creates the tides on earth. The moon's gravity tends to elongate and compress the earth. Rock yields less to this force than water, so we notice tidal bulges of the oceans.

The fantastic tidal force near a black hole also tends to stretch any object in the radial direction and compress it in the other directions. Picture yourself floating in space, your feet pointing to a black hole and your head away from it. Tidal forces would stretch you from head to foot, and crush you from the sides.

Identical forces would be felt by a rocket or any other object. The forces at the event horizon of a black hole a few times more massive than the sun would certainly be enough to kill a person, and possibly enough to destroy a sizable object of any known material. No one could survive approaching, much less entering, a typical-size black hole.

Black holes come in different sizes. The size of the black hole (or, more exactly, of its boundary, the event horizon) depends on the mass of the object that created it. Ironically, the tidal forces at the event horizon are *less* for more massive black holes.

According to general relativity, the tidal forces at the event horizon decrease in inverse proportion to the square of the black hole's mass. It has been estimated that a human body could withstand the tidal forces existing at the horizon of a black hole with a mass about a thousand times that of the sun. No known star is that big, but it is suspected that there are black holes much bigger yet.

In 1987 astronomers Douglas Richstone and Alan Dressler reported evidence that massive black holes may exist in the center of the Andromeda galaxy and its satellite galaxy, M32. They found that stars near the galaxies' centers were orbiting much faster than expected. In the case of the Andromeda galaxy, this could be explained if the stars were orbiting an unseen, relatively compact object with a mass about 70 million times that of the sun. Among known and theoretical objects, that could only be a black hole. Other, more indirect evidence suggests that a similar black hole may exist in the center of our own galaxy. The tidal forces at the event horizon of such a black hole would be mild—about 5 billion times less than that of a 1000-solar-mass hole. A person could easily survive the tidal forces outside a galactic black hole and for some distance inside.

At the center of a black hole is a "singularity," a point of infinite compression and infinite curvature of space-time. Any object that crosses the event horizon is pulled to the singularity. For an observer, reaching the singularity is the end, no matter what. No body or device can withstand infinite tidal forces.

The time it takes to reach the singularity depends on the size of the black hole. It works out to 1.54×10^{-5} seconds times the mass of the black hole divided by the mass of the sun. (This is the time measured by the infalling observer. To other observers, the time interval is different. In the frame of reference of an observer at rest far from the black hole, the fall takes—literally—forever. This is another effect of the profound distortion of time and space around a black hole.)

For a typical black hole of two solar masses, the travel time from event horizon to singularity would be about 3×10^{-5} second. For a black hole of 1000 solar masses, the maximum time of fall would be 0.0154 seconds. In either case, an observer would be dead by the time he crossed the event horizon.

However, for a 70-million-solar-mass black hole, such as might exist in the heart of the Andromeda galaxy, the time is 1100 seconds (18 minutes). Tidal forces would remain bearable for nearly all of the 18-minute fall to the singularity. Certain death would come only in the last split second.

The ultimate fate of someone entering a black hole is surrealistically gruesome. In the last moments before hitting the singularity, tidal forces would increase without limit. Bone and muscle would give way, followed by cellular and atomic structure. You would be spun into a spaghetti strand of ever-increasing length and ever-decreasing diameter. The strand would become thinner than the finest thread while stretching to infinite length (the radius of a black hole is infinite from the inside). The final volume would be zero. A human body that enters a black hole is transformed into Euclid's ideal line.

(The infalling observer's view of all this would probably be disappointing. You would want, I guess, to see the singularity, or at least to see all the previously consumed objects that have been deformed into radial needles of zero volume. Unfortunately, the light from all previously consumed objects, including the stars that formed the black hole, can never reach a later infalling observer. You could only see objects that crossed the event horizon after you. Like Brahma, the singularity is visible to no observer until he becomes part of it.)

The experiment could not be popular, but its conceivability has bearing on the "reality" of black-hole interiors. There would be time for an infalling observer to take photographs, do experiments, write about the experience in a diary. To the observer, there would be no doubt about the reality of the experience.

The catch is that there is no possibility whatsoever of the observer communicating his experiences to us on the outside. The experience cannot be incorporated into the body of shared human experience. Does this make a difference? If you think it does, suppose the earth falls into this galactic black hole. For 18 minutes, *everyone* would be conscious of being inside a black hole.

One feels strongly that all this demonstrates that black-hole interiors *are* real (provided general relativity is right). There is a big difference between a hypothesis that no one could confirm, no matter what (like Poincaré's nocturnal doubling), and one that is only very difficult—even suicidal—to test (the astrophysics of black-hole interiors). The realm of science is testable hypotheses—hypotheses that somehow "make a difference." Poincaré's doubling is a phantasm because there would be no difference. In contrast, if you entered a black hole, something would happen, and that something would either confirm or refute general relativity.

Other Minds

Cognitive science, the study of mind, deals in many untestable entities. The venerable "other minds" problem of philosophers asks how we know that other people have thoughts and feelings as we do. Everyone else *could* be a robotlike being programmed to talk and react, but feeling nothing. What can you do to demonstrate that this is not so?

The "other minds" problem can be formulated as a Poincaré-esque thought experiment. Suppose that last night everyone except you lost their soul/consciousness/mind. They still act the same way, but the internal dialogue, so to speak, is completely gone. Is there any way of telling that this has happened? (Or assume that half the people in the world have souls and the other half don't. How do you tell who does and who doesn't?)

Certainly other people talk about their loves, hates, pains, and joys. That proves nothing. We have to assume that all the observed diversity of human behavior is within the ken of the unconscious automatons. If other people's consciousness is an illusion, it is a good illusion.

What you want is some clever question that would catch the alleged robots off guard and reveal their lack of true emotion. You could say that the fact that other people have thought up and discussed the "other minds" problem is proof that they have minds. Robotlike beings would not care, or even suspect, that there might be true consciousness. This fails to give the hypothetical robots enough credit.

There is inductive reason for belief in other minds. In many, many ways, each of us learns that we are similar to the five billion other members of the human race. Since each of us (presumably) knows he has a mind, it is natural to project that attribute onto everyone else. This is shaky induction, for it is extrapolating from just one known mind. Hence the question about finding an objective test.

The test would have to be a way of "getting inside someone else's head" and feeling what the other person feels (or doesn't feel). ESP, if it exists, might allow that; so might some sort of futuristic brain experiment in which one person's brain is artificially linked to another's. Even these exotic measures might not eliminate doubt entirely. It still could be that you are the only person with consciousness and are responding to "brain waves," "auras," or "vibrations" produced by the automatonlike brains of others.

Most philosophers concede that whether others experience consciousness is strictly unknowable. Some take this one step further and argue that consciousness and perfect simulation of consciousness are the same. Here most people object. You probably feel that there *is* a difference between consciousness and the lack thereof, even while admitting that no possible observation or experiment would establish it. Is this a rational objection?

Nocturnal Doubling of Pleasure/Pain

A clever recent twist on Poincaré's thought experiment asks what would happen if everyone's sensations of pleasure and pain doubled overnight. It is considerably less clear than in the original that this is meaningless, though some of the same reasoning applies.

In 1911 economic theorist Stanley Jevons wrote:

> . . . there is never, in any single instance, an attempt to compare the amount of feeling in one mind with that in another. The susceptibility of one mind may, for what we know, be a thousand times greater than that of another. But, provided that the susceptibility was different in a like ratio in all directions, we should never be able to discover the difference. Every mind is thus inscrutable to every other mind, and no common denominator of feeling seems to be possible.

Jevons is saying that, just possibly, your friends' sensations are a thousand times greater than your own. Or a thousand times less. Consider, then, this thought experiment:

During the night, pleasure/pain doubles—meaning that any particular stimulus, such as a slice of pecan pie, sexual climax, or a bee sting, henceforth causes twice the pleasure or pain that it did before. It must be stipulated that only the *subjective* sensations are doubled. Pleasure and pain are associated with certain measurable brain activities. Were the levels of endorphins (brain chemicals associated with some types of pleasure) higher, or were there a measurably increased electrical activity of the C fibers (which have been associated with pain), the change would obviously be detectable to a neurologist. Not so obvious is whether a subjective doubling would be detectable.

The first thing to ask is if preferences (the choices you would make when free to make them) would be any different. Presumably not, since preferences seem to be based only on relative degrees of pleasure and pain.

Philosopher Roy A. Sorensen concluded that a doubling of preferences would not be detectable. You walk into an ice-cream shop the day after the change. The shop has thirty flavors of ice cream, of which twenty-nine you like in varying degrees, and one (licorice) you detest. Since pleasure/pain has doubled, the licorice ice cream is twice as loathsome now. Of course, you wouldn't have ordered licorice even before the doubling. You would have ordered your favorite flavor, unless the desire for novelty was strong enough to override that preference and lead you to choose another.

Now, after the doubling, you do the same thing. The favorite flavor beats out the second favorite by twice as big a margin. The pleasure of novelty is twice as great too, and you might elect to try a new flavor rather than your favorite, but only if you would have done so even had the doubling not occurred. In general, diners would make the same selections from menus; so would condemned criminals given their choice of execution method; it would neither help nor hurt any television show's market share.

George Schlesinger (who argued that Poincaré's physical doubling would be detectable) claimed that a doubling of preferences would be detectable through indiscernible preferences. He argued, in effect, like this: Suppose that if given a choice between a bee sting and a wasp sting you cannot decide because each gives you almost precisely the same degree of pain. After pleasure/pain is doubled, however, there is more "room" between them on your personal scale of preferences, and you can see that the bee sting actually hurts less. You might even be able to find some pain that fits between them. Maybe it would be evident that you prefer a tax audit to a bee sting, and a wasp sting to a tax audit. Sorensen countered that this is no different from saying that a nocturnal doubling of length could be detected by reexamining pencils that had previously seemed to be exactly the same length!

The pleasure principle of Freudian psychology claims that we always choose to do what is most pleasurable (whether for the moment or in the foreseeable long run). If that is true, it should not matter that the most pleasurable thing is twice as pleasurable. On the television quiz show *Jeopardy,* contestants answer questions on a game board for designated amounts of money. After a commercial break, they start with a new board called "Double Jeopardy," where each question is worth twice as much. Obviously, the strategy for "Double Jeopardy" is exactly the same, even though you win twice as much money. This agrees with a basic tenet of decision theory, which says that multiplying "utility" (a numerical measure of how much one desires or does not desire a particular outcome) by two or any positive factor makes no difference. What was preferred before will still be preferred.

Some have pointed to the "pleasure center" experiments of James Olds and Peter Milner as proof that a doubling of pleasure/pain would be noticeable. In the early 1950s, Olds and Milner implanted electrodes of silver wire in the brains of rats to see how electrical stimulation of the brain could influence behavior. They were looking for hypothetical "avoidance centers" where stimulation would teach rats to avoid behaviors. The rats roamed freely across a table. When a rat approached one corner, the experimenters applied an electrical impulse (5 to 100 microamperes for half a second) to the embedded electrode.

Olds and Milner found very strong avoidance centers. Stimulating these parts of the brain as a rat approached the forbidden corner caused it to turn around and flee. A single such experience taught the rat to steer clear of the corner indefinitely. Then came one of the accidents that are so much a part of the history of science. As one rat approached the corner and received its electrical stimulation, it stopped. It moved a few steps further toward the corner, and stood fast. When moved away from the corner, it tried to return. Olds and Milner examined the rat more carefully and found that its electrode had been implanted in a slightly different part of the brain. This part of the brain had the opposite function from the avoidance centers.

This new site came to be called a "reward" or "pleasure" center. Conversely, the avoidance regions were guessed to be sites of pain. Rats would readily learn their way through mazes to receive stimulation of the pleasure centers. Rats that were allowed to stimulate their pleasure centers by pressing a lever soon did nothing else. They pressed the levers a hundred times a minute until they collapsed of exhaustion; after brief sleep, they immediately began again.

The identification of the sites as pleasure and pain centers was tentative. Olds and Milner faced a rodent "other minds" problem: Do rats experience pleasure and pain as we do, or are they virtual automatons? Later experiments were done on human volunteers. The sensation produced by stimulating a pleasure center was pleasurable (but not nearly so compelling as it seems to be with the rats). Psychologists have identified dozens of distinct pleasure centers of the brain, associated with sex, food, thirst, and other basic desires.

The feeling of some is that if pleasure/pain doubled, we would all be like Olds and Milner's rats, immersed in a nonstop orgy of pleasure. In Olds and Milner's experiments, however, the pleasure of just one action (that of pressing the lever in the cage) increased unilaterally. That made the rat's preferences different. If every action increased in pleasurableness equally, the situation would be much more similar to Poincaré's original situation.

Suppose you are in the ice-cream shop eating a dish of your favorite flavor of ice cream. It tastes twice as good as it ever did. Does that mean you would gorge yourself on it? The stomachache you would get from eating too much ice cream is now twice as unpleasant; you are watching your fat intake and hate to go off your diet twice as much. Also, you are constantly making a choice between eating and doing something else. Eating has high priority when you are hungry; when sated, it has low priority. All the other things you could be doing instead of having a second dish of ice cream would be twice as attractive too.

Even if people acted the same, wouldn't they still be aware of the doubling? You might detect the change by comparing your current pleasure and pain with memories of past experiences. The fact that we make statements like "This is the best pecan pie I ever tasted" shows that we carry memories of past pleasures and can use them to gauge current ones.

I tend to agree with this and yet am unsure just how it is different from a statement like "From the looks of it, the Sears Tower is the tallest building I've ever seen." The height of a building is judged in one of two ways. The objective way is to consult published accounts of its height. A guidebook lists the height of the Sears Tower as 1454 feet; you can compare this with the heights of other tall buildings you have seen and conclude that the Sears Tower is the tallest. How would you compare pleasure/pain objectively? It could only be through published accounts of past preferences (such as results of a wine tasting comparing vintages). Those accounts would compare (old) degrees of pleasure/pain to other (old) pleasures and pains and would therefore be of no help at all. It would be like trying to gauge the height of a doubled Sears Tower with doubled yardsticks.

The subjective way of gauging a building's height is by comparing it with nearby buildings, the angle you must turn your head to see its top (in effect, a comparison with your own height), etc. It is likely that some of our experience of pleasure and pain entails a comparison with contemporaneous pleasure and pain. (The best meal you ever tasted was what you had after spending all that time at summer camp/in prison/on a raft with no food; hitting your head against the wall feels good when you stop; euphoria follows the pain of giving birth.) With all pleasures and pains doubled, such comparisons would fail to detect the change.

If you feel that memories would give away the change, let the change occur gradually (even over a period of centuries). Is there anything Plato could have written that would convince us that the Hellenes felt twice the pleasure and pain that we do in the effete twentieth century?

More difficult to refute is the claim of some philosophers that stress would be greater. What if you walk into a foreign gambling casino, find a green roulette chip on the floor, and use it to place a bet on lucky number 7? It's a 100-smackeroo chip, which you figure to be worth $2 American. Just after you irretrievably place the bet, a friend tells you that you made a mistake about the exchange rate, that the chip is actually worth $2000. You will either lose the $2000 or win $72,000. The ratios of the wins and losses are exactly the same, but still, wouldn't you be more jittery with the higher stakes? It appears there would be greater stress in a world with doubled pleasure/pain. There would always be twice as much to gain and twice as much to lose.

One response is that, yes, there would be twice as much stress, because stress is a form of pain and is doubled. The ratios would still work out the same. On the other hand, this stress might be manifested in a higher rate of ulcers, increased use of tranquilizers, greater suicide rate, etc.—objective changes.

Arguably, sadists and masochists would detect a doubling of pleasure/pain. Not only would sadists derive twice as much pleasure from inflicting a given amount of pain, but any given cruelty would cause twice as much pain. A sadistic act would cause twice the pain, and thus four times the pleasure. Comparable reasoning applies to masochists: Their pain is doubled, but the pleasure per "unit" of pain is doubled, meaning four times the pleasure.

What spoils this ingenious idea is that no one, including sadists, really knows other people's pleasure and pain (or even if they have minds at all). It is the counterpart of Schlesinger's treatment of physical doubling. How would the sadist know that pain has doubled?

Is Reality Unique?

All these examples demonstrate that there are many wildly different hypotheses that are compatible with experience—an *infinite* number, Poincaré said. The scientific method is powerless to rule out these alternative hypotheses. Can we say that a hypothesis like nocturnal doubling is true or false?

Poincaré felt that some of these irrefutable hypotheses are easier to work with but not necessarily truer. Poincaré's view is distressing to many. Rather than one reality, there are many; you are free to take your pick.

"A reality completely independent of the spirit that conceives it, sees it or feels, is an impossibility," Poincaré wrote. "A world so external as that, even if it existed, would be forever inaccessible to us. What we call 'objective reality' is, strictly speaking, that which is common to several thinking beings and might be common to all; this common part, we shall see, can only be the harmony expressed by mathematical laws."

PART TWO

INTERLUDE

The Puzzles of John H. Watson, M.D.

In solving a problem of this sort, the grand thing is to be able to reason backward.
—SHERLOCK HOLMES, *A Study in Scarlet*

SEVERAL YEARS had passed since Sherlock Holmes retired to the placid life of a beekeeper on the Sussex Downs. His missive (the first) read simply: "The bucolic life does not altogether suit me. How I hunger for mental stimulation! Can you arrange a visit?" I canceled the few appointments on my schedule and booked passage on a train south the next day.

As we shared a late supper that night, I remarked that I had been analyzing our old adventures at my club. "Holmes, I think you are one of the most misunderstood men in Britain. All think your renown is due to the difficulty of your cases. I believe that what made the tales popular is that the solutions were so *simple.*"

Holmes put his fingertips together, his face now betraying amusement. "You believe the British public wants to hear of simpleminded detection?"

Feeling pleasantly articulate from the port, I continued: "The public likes a case where the solution, once stated, is pellucidly obvious and self-evident . . . however difficult it may have been to come up with that solution. Only when the solution is so obviously correct can the reader kick himself for not having thought of it himself."

"*All* problems are easy in retrospect," Sherlock Holmes countered. "It is like solving a labyrinth by going backward from the goal."

"No," I objected. "I must differ. *Some* labyrinths are just as difficult when you start from the goal. There are many problems whose solutions are as abstruse as the problems themselves. If you had gone around hanging men on dreary ballistics and fingerprints, the way Scotland Yard does most of the time, my accounts of your exploits would not have found one-tenth the audience they did. The public wants a readily understandable solution."

"An interesting point," Holmes said dreamily. "I subscribe to several esoteric journals to ease the tedium of the apiarist's life. I was reading in one of them that William Shanks, a mathematician of our fair island, has recently computed pi to 707 decimal places. It took him twenty years. His result filled a whole page with quite senseless, random numbers. Should anyone doubt Mr. Shanks's result, he would have to budget an equal amount of time and duplicate his work. In that case also, verifying the answer would be precisely as difficult as coming up with the answer in the first place —the very antithesis of an 'obvious' solution."[1]

"Precisely. I was telling this to a school chum in London, and he said it's like this riddle: What common English word starts and ends with the letters UND? It's difficult to think of the word, but once you think of it, you can't doubt you've got the right answer. It's no fair checking in the dictionary, of course."

Holmes wrinkled his forehead at this but said nothing.

"I told several acquaintances at the club that I was visiting you and wanted to pass the time with a few riddles. They gave me some good, tough puzzles. They're all the sort you seem to specialize in, where the answer is obvious once you see it. That way, you can mull

---

[1] Unknown to Holmes's generation, Shanks made an error on the 528th decimal place. All the digits after that were wrong.

them over for weeks if need be, and I needn't be here to tell you you're correct."

"Weeks? I hardly think so."

A Test of Ingenuity

After the meal, I directed Holmes into the guest room next to mine. It had known little use under Holmes's tenancy and was sparsely furnished. That afternoon, I had removed the bed and chair, leaving it quite bare.

Hanging from the ceiling were two lengths of string, each six feet long. The strings were ten feet apart. As the room was also ten feet high, the lower ends of the strings hung four feet above the floor.

The only other feature of the room was a eclectic assortment of objects laid out on the floor. There was a Swiss Army knife, a firecracker, a small vial of ether, a twenty-five-pound block of ice, and a tortoiseshell kitten. The ice was in a pan to avoid harm to the Indian rug.

"I concede, Watson. What are you up to?" Holmes asked.

"The problem," I said, "is to tie the two ends of the string together. You will notice that the distance between the hanging strings is about four feet greater than your arm span. While holding one string, you cannot touch any part of the other string. All you are allowed to use in your solution are the Swiss Army knife, the firecracker, the ether, the ice, and—*or* the kitten. You may not use the curtain rods, wallpaper, the carpeting, or anything else in this room. That includes clothing and objects on your person."

Holmes's eyes inspected the floor and ceiling minutely. "The ladder you used to hang the string had a loose third rung."

Ignoring this, I continued: "I have tied the strings to the ceiling fixtures with simple slip knots. They will not support your weight. As a hint, I can only remind you that it's one of your infuriating puzzles where the solution, difficult though it may be to deduce, is absurdly simple after you see it."

Gas, Water, and Electricity

Holmes spent a few moments in silent introspection. Then he asked, "The second puzzle?"

"Your next problem," I said, "is one that was recently written up in the papers by Henry Ernest Dudeney. I spent some time mulling this over, only to hear that it has no solution at all. Then I thought it over some more, and concluded that there is a solution."

I showed Holmes the diagram that I had clipped from the paper. "There are three houses and three utility companies—gas, water, and electricity. Each company desires to lay pipes or cable to each house without any pipe or cable intersecting the path of any other. The paths of the pipes can bend and may be wastefully indirect; they cannot cross."

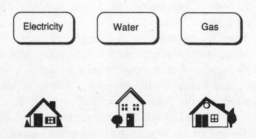

Holmes scarcely glanced at the clipping. "I am quite familiar with that puzzle, Watson. It is older than electric lighting or even gas lighting. Less modernistic versions spoke of paths to a dovecote, a well, and a haystack. I can assure you you are mistaken if you think you have a solution. It can't be done."

"I think you will agree with me that there *is* a solution nonetheless."

Holmes sighed tolerantly. "I am aware of the *unfair* solutions that have been offered. One pipe could go through one house. The water pipe could be concentrically embedded in the gas pipe. I confess a certain admiration for the cleverness of these schemes, but I trust you appreciate that they are cheats. They violate the essentially topological spirit of the original puzzle. The houses and utilities should be regarded as dimensionless points, the pipes as curves of zero breadth."

"I could not agree with you more. There is a solution that is true to this topological spirit of which you speak."

The Company Grapevine

My third puzzle went like this: "A certain large business firm has 1000 employees, and a peculiar method of terminating them," I began. "Never is anyone told that he is fired. Each employee scheduled for termination is allowed to deduce his impending fate and resign rather than be fired.

"All the company's employees live in constant fear of losing their jobs. Rumors of impending firings spread instantaneously through the company. This rumor mill is completely accurate. There is such a steady stream of firings that no one invents falsehoods out of malice or boredom. Whenever someone is scheduled to be let go, everyone in the company knows about it except for the unfortunate person himself. He is literally the last to know. No one has the grit to tell anyone that he is going to be fired, and everyone has learned through constant practice to act precisely as if nothing is wrong when in the company of a doomed fellow worker.

"This environment of gossip and duplicity has made the logical powers of the employees all the keener. Each man lies awake in bed each night thinking over what he heard and what he didn't hear, hashing over all possible hypotheses concerning his position in the company. No nuance, no slight goes unrecognized or unpondered. All the employees being quite bright (and quite paranoid), no one fails to see all logical implications of any action. If an employee deduces that he is to be fired, he hands in his resignation the first thing the next morning.

"One day the company was acquired by a larger firm. The manager of the larger company called a meeting of all the employees and said, 'It's time somebody trimmed the fat over here. Heads will roll!' The manager did not say who was to be fired. He did not even say how many people were to go. As always, there were no secrets to the company grapevine. Immediately after the meeting, the grapevine learned who was to go. What happened next?"

"What do you mean, what happened?" Holmes asked.

"A rather beautiful deduction about the firings is possible. The puzzle is to see what that deduction is."

"There isn't enough information!"

"The appeal of this little conundrum, Holmes, is how much can be deduced from a minimum of information."

Holmes appeared to toy with several ideas and reject them all. "I suppose some of the wretches could figure out what was up from the way other people acted."

"No, no, you miss the point. They are consummate actors all, and so spineless that they would not tell their best friend of his fate."

"I have noted that the pupils of the eye frequently betray the most practiced liar—"

"I said nothing about pupils, so it can't be relevant."

"The employees cannot get together and pool their knowledge?"

"Not other than in the company rumor mill as I have described it. No one, under any circumstances, tells anyone that he is to be fired, or allows him to be told by a third party outside the company."

"Anonymous letters?"

"Not permitted."

The Graveyard Riddle

"Speaking of anonymous letters: A man gets an unsigned letter telling him to go to the local graveyard at midnight. He does not generally pay attention to such things, but complies out of curiosity. It is a deathly still night, lighted by a thin crescent moon. The man stations himself in front of his family's ancestral crypt. The man is about to leave when he hears scraping footsteps. He yells out, but no one answers. The next morning, the caretaker finds the man dead in front of the crypt, a hideous grin on his face.

"Did the man vote for Teddy Roosevelt in the 1904 U.S. presidential election?"

"Well!" Holmes said with greater enthusiasm. "At last a problem open to *logical* solution!"

A Surveyor's Quandary

I next produced a set of three cardboard figures from my doctor's bag: a triangle, a square with one quarter missing, and an entire square. "In a certain part of the American desert, there were three landowners I will call Smith, Jones, and Robinson. Smith had three sons, Jones had four, and Robinson had five. Americans being very democratic, they divide their estates equally among all their heirs.

"Smith's property was in the form of a regular triangle. He did not want to favor any of his three sons, so he asked the county surveyor to divide the land into three tracts of exactly the same size and shape. That the surveyor was able to do." With a pen I sketched the division on the cardboard triangle.

"Jones, with four sons, had this L-shaped property, three-quarters of a square. After much deliberation, the surveyor divided it into four parcels, each the same size and shape.

"Finally, Robinson, with five sons, had a perfectly square property. He asked the surveyor to divide it into five identical pieces.

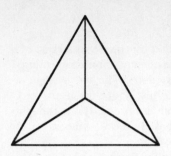

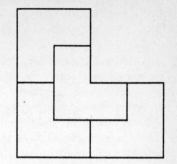

The surveyor found the problem quite intractable. He could not put it aside, and he neglected his other work. He ended up tearing most of his hair out, and they had to feed him with a spoon. Your final puzzle is to divide a square into five pieces, all exactly the same size and shape. It *is* possible, but I warn you that there is only one way of doing it.

"I hope these diversions occupy you. I plan to retire for the evening. Please don't stay up all night. If you do, I would appreciate not being woken when you hit on the answers. You shan't need my verification anyway. The correct solutions will be as plain as the nose on your face."

As I left Holmes, he was seated at the deal-topped table, scribbling notes and ignoring me.

Solutions

I had bad dreams, occasioned, I think, by the weird strains of a violin playing through the night. I arose at eight the next morning. I first peeked into the empty room next door. The two strings attached to the ceiling now hung knotted just above my head.

I was relieved to find the kitten fit and unharmed. It was thoughtless of me to employ it as misdirection! Holmes's compassion for animals was not excessive, I recalled.

I found Holmes lying on the couch in the drawing room in a blue haze of tobacco smoke. He was still wearing last night's clothes. "All but one were quite trivial," he announced.

"Really?" I sat down at the table. A thousand mad diagrams dissected a thousand squares into crazy quilts. At the top of one sheet of paper was the desperate neologism UNDACHSHUND, heavily crossed out. With difficulty I refrained from commenting. Below it was the correct answer.

"Which did you get first?"

"I got the UND puzzle *last.*"

"That was supposed to be the easiest."

"So I thought as well," Holmes conceded. "The riddle is hard because there is no systematic way of solving it. If inspiration fails, the best you can do is to run through all the possible combinations of letters beginning and ending with UND.

"Look here," Holmes said, proffering a sheet of paper covered with letters. "Since UND is not a word, nor UNDUND, we have UNDAUND, UNDBUND, UNDCUND, and so on through UNDZUND. If none of these 26 seven-letter combinations are common English words (they aren't, I quickly found), you must run through the eight-letter groups: UNDAAUND, UNDABUND . . . UNDZZUND. This time, however, there are 26 times 26 combinations. That's 676 all told."

"And you still won't have the answer," I added.

"No. As you check longer words, the number of combinations goes up in a geometric progression. The correct answer, you will observe, is long enough so that one would have to check *millions* of letter combinations to stumble on it. That is why I find this problem unfair. No one could ever solve this problem logically; it's too taxing."

"How did you get the answer?"

"A lucky guess. The so-called subconscious mind. Either of

which is unsatisfying. I had hoped to deduce it logically. One moment all was darkness; the next, UNDERGROUND popped into my head."[2]

"There were perhaps other instances of a lucky guess triumphing over deduction?" I suggested.

"Tying the string? To a degree. By now I recognize a red herring when I see one, Watson. You must not be too upset when I tell you I suspected right off that the more fanciful of your set of objects were quite likely irrelevant. The clever thing on your part was that the solution depends not on a specific object but on any of them.

"I used the Swiss Army knife. The bottle of ether would work as well, and the firecracker or ice might serve the purpose. The cat would wriggle—I suppose the ether could remedy that. I tied the knife to a string and set it to swinging. Then I grabbed the other string, caught the knife, and tied the strings in a graceful parabola. It is simplicity itself—in retrospect."

"A graceful catenary," I corrected.

"I commend you on bringing the gas, water, and electricity puzzle to my renewed attention," Holmes said. "I gather this is what you had in mind? Your topological solution?" He produced a neat pencil diagram of my solution.

Holmes explained: "The riddle is given as a problem on a plane. That the earth is actually a sphere makes no difference. Any network of points and lines on a sphere is equivalent to a network on a plane, since the sphere may be 'punctured' at the antipodes and deformed to a plane. The paths of the pipes need not cross on certain other topological surfaces. The problem is soluble on a Möbius strip or on a torus—a doughnut shape, with a hole in it. Any natural tunnel renders the earth a torus. 'Natural bridge' or 'window rock' formations, caverns and sea grottoes with two openings, blowholes, prairie-dog burrows—any will do. The tunnel is, in effect, a free crossing. If you have two pipes that must otherwise cross, you may route one pipe through the tunnel, and the other over the mountain, so to speak.

"The torus hole must be within the network of pipes. It is presumptuous to assume a priori that any tunnel or burrow exists in the vicinity of the houses and the utility companies. This threw me for a while. Then I realized that if the mountain won't come to

---

[2] An answer not likely to be found in the dictionaries of Watson's day is UNDERFUND.

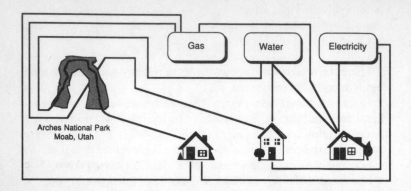

Mohammed, Mohammed may go to the mountain. It is proper to assume that the puzzle takes place on this earth, and there are many natural tunnels in the earth. The utility companies could snake three of their pipes out to the nearest such tunnel or burrow and then back to the houses."

"I trust you got the one about the company rumor mill," I said. "That was pure deduction, I should say."

"It was most singular. The answer is pure deduction, and yet I am not at all sure that I deduced. I fear it was another fortunate guess."

"A guess?"

"The Medicis were said to have had a slow poison which proved fatal only after as many days as it had been cured in the sun. If one wanted a collateral heir or indiscreet mistress to die in fifteen days, one administered a preparation that had baked in the Florentine sun for fifteen days. The formula has been lost to us—"

"As a medical man, I say that's a fairy tale. What about the riddle?"

"I mentioned the Medicis' alleged poison only because thinking of it (quite by chance) led me to the solution. Everyone in the company scheduled to be fired will deduce it and resign on the same day. It will take as many days to arrive at this deduction as there are people to be fired. If the company is letting 79 people go, then on the 79th day after the announcement, all 79 people will resign."

"And how did you come to that conclusion?" I asked.

"It is a wonderful—a monstrous—bit of logic. I simplified the puzzle by supposing that just one person is to be fired. The rumor mill learns his identity, and everyone in the company except that person knows the situation. That night, the doomed man is tossing and turning in bed. He knows that someone is going to be fired. Isn't it peculiar that he hasn't heard who? The company grapevine is so efficient . . . The only possible conclusion is that he *and he alone* is to be fired. Were he one of a group of persons to be fired, he would have learned the names of the others. Consequently, this single unlucky man must resign the next morning. It is the only logical possibility. *If* just one person is to be fired, that is precisely what does happen.

"Allow instead that there are two persons to be fired. Thanks to the grapevine, everyone learns the name of at least one person to be fired and sleeps soundly the first night. Each employee can suppose that the one-person scenario I have just outlined is taking place. The *second* night after the announcement, the two persons to be fired are stricken with insomnia. Each thinks, 'It's too bad about so-and-so getting the sack. What I can't figure is, why didn't he resign this morning?' All employees being perfectly logical and having ample time to consider the implications of actions, so-and-so could have failed to resign *only* if he knew the name of another employee to be terminated. Each of the two employees must conclude that that other employee can only be himself. Both employees must resign on the second morning after the announcement.

"Then it all fell into place. If three persons are to be fired, each can deduce his fate from the facts that no one resigned on the first and second mornings and that each knows only two persons who are scheduled to be fired. It makes no difference how many persons are involved, just as long as all can trust in the unflinching logic of their fellows. After 999 days with no resignations, all 1000 employees would spend a sleepless night concluding that the entire work force is to be released."

"And the man in the graveyard?"

"I told you, Watson, that was the easy one."

"To you perhaps. I'm not sure there is an objective way of saying what's easy and what's difficult."

"I suppose you're right. In any event, the answer is of course no, the man didn't vote for Roosevelt. The solution depends on realizing that the aforesaid crescent moon cannot be visible in the middle of the night. Not from most parts of the world, anyway. It is shocking how many of the so-called educated class are unaware of this elemental fact known to every goatherd. The exception is in the polar regions, where the sun—and nearby crescent moon—can be

visible all through a twenty-four hour cycle. Therefore, if the man
lives in the United States at all, he must live in Alaska, near or
above the Arctic Circle. The citizens of the territory of Alaska are
not permitted to vote for President. Whatever his politics, the man
did not vote for Roosevelt."[3]

"Congratulations, Holmes," I said. "Well then, it must be the
land-division puzzle that baffles you."

Holmes nodded. "It is that which is responsible for keeping me
up all night. I feel this puzzle is qualitatively different from the
others. With the others, the number of conceivable solutions is in
some sense limited. But just as there are an infinity of lines in a
plane, so are the possible dissections of a plane figure beyond number. Not only did I fail, I did not even see how to *start.*"

"Are you ready to concede defeat?"

"Yes. Show me how to do it."

"I have given it some thought, and think it best that I inform you
by letter, once I am safely back in London."

"Why?"

"You will not be happy."

"Watson, tell me at once!"

Only upon arriving home the next day did I post this diagram to
Holmes:

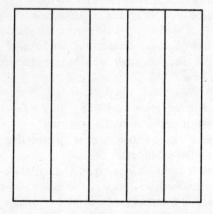

[3] The puzzle may be updated to ask about any presidential race up through 1956 (Eisen-
hower vs. Stevenson). Alaska became a state in 1959, qualifying its citizens to vote in the 1960
election. If you ask whether the man voted for Kennedy or Nixon in 1960, there is not enough
information to answer.

5

DEDUCTION

The Paradox
of the Heap

P UZZLES AND PARADOXES are subtly related. In a puzzle just one of many conceivable hypotheses avoids creating a contradiction. That single hypothesis is the puzzle's solution. In a paradox, no hypothesis at all is tenable.

Like raw oysters, logic puzzles are an idiosyncratic taste. Some find them challenging or amusing; others, annoying. An important question is whether there is any general method for solving logic problems. Is there a cut-and-dried procedure, a trick, a recipe that anyone can learn and use to tackle any logic problem? If there is, it would be invaluable in the scientific realm and elsewhere.

In practice, logic is a blend of step-by-step deduction and exhaustive search of possible hypotheses. The first approach is illustrated by a set of classical paradoxes.

Theseus' Ship

When Theseus returned to Athens after slaying the Minotaur, his ship "was preserved by the Athenians down even to the time of Demetrius Phalereus, for they took away the old planks as they decayed, putting in new and stronger timber in their place," wrote Plutarch. "This ship became a standing example among philosophers, for the logical question as to things that grow; one side holding that the ship remained the same, and the other contending that it was not the same."

Everyone concurs that replacing a plank on a ship does not change its identity. It is still the same ship after the replacement. Replacing still another plank on the repaired ship should not make a difference either. At some point, perhaps, Theseus' ship contained not a single plank of the original ship. Then surely the Athenians were deluding themselves in calling it Theseus' ship. Had the ship *not* been preserved, and had the latter-day Athenians constructed a ship directly from those same nonoriginal planks, no one would dream of calling it anything but a good replica of Theseus' ship.

Minor paradoxes of this type were popular with the ancient Greeks. A millet seed falling makes no sound, said Zeno. Then how can a bushelful of millet seeds make a sound when it falls, if it contains nothing but millet seeds? Of kindred spirit is the "paradox of the heap": Whenever you remove a grain of sand from a heap of sand, you still have a heap of sand. Picture a heap of sand, and take away a single grain. Is there any possible way, based on your past experience, that removing a grain of sand could leave you with something other than a heap of sand? Of course not. Then start with a heap of sand and subtract the grains one by one. Eventually the heap dwindles to a single grain. It still must be a heap! Then remove the lone grain, leaving nothing. The nothing still must be a heap!

Sure, the way out is to set a minimum size for a heap. "A heap must have at least 1000 grains, so the rule should be: 'Removing a grain of sand from a heap containing at least 1001 grains leaves a heap.'" *That* leaves a bad taste in your mouth even as you utter it. Doesn't it miss the point? A word like "heap" is supposed to be vague.

A modern counterpart is Wang's paradox (after Hao Wang). Wang claims that if a number x is small, then $x + 1$ is also small. Does everyone agree that 0 is a small number? Okay; then 1 (0 + 1)

is small. And 2 (1 + 1) is small. And 3 (2 + 1) is small. And so on
. . . *every* number is small, which is ludicrous.

Sorites

A *sorites* (pronounced *suh-RYE-teez)* is a chain of linked syllogisms—the form of argument in which the predicate of each statement is the subject of the next. In other words, it's this:

> All ravens are crows;
> all crows are birds;
> all birds are animals;
> all animals need oxygen.

The premises of a sorites join up and lead to an obvious conclusion ("all ravens need oxygen"). Recognizing sorites (the plural is
spelled the same way) is the key to many logic puzzles. The com-
pany-grapevine solution of the previous chapter is an elaborate
sorites.

The sorites is named for the Greek word for heap, since it is the
form of reasoning used (fallaciously) in the paradox of the heap:

if x is a heap, then x less 1 grain is a heap;
if x less 1 grain is a heap, then x less 2 grains is a heap;
if x less 2 grains is a heap, then x less 3 grains is a heap;
if x less 3 grains is a heap, then x less 4 grains is a heap;

> .
> .
> .

if x less 12,882,902 grains is a heap, then x less 12,882,903 grains is a
heap.

Here the number of distinct logical steps may be millions.

Sorites paradoxes are possibly the simplest paradoxes of deduction. None qualifies as baffling. All derive from the way the slight
inaccuracy of a premise can accumulate when the premise is applied
over and over. What charm the sorites paradoxes have is that they
use (abuse) a very common and important type of reasoning. Most
of what we know or believe is through sorites.

One day you see a raven that neither you nor any ornithologist
has ever seen before. Even so, you know a lot about that raven. You
know (or have strong reason to believe) that it is warm-blooded,
that it has bones underneath the feathers and skin, that it was
hatched from an egg, that it needs water, oxygen, and food to sur-

vive, etc., etc. You know all this neither through direct experience nor through being told it explicitly. Did you ever put a raven (much less *that* particular raven) in a room full of pure nitrogen? Did you ever read in a book the flat-out statement: "All ravens have bones"? You know these facts about the raven through sorites you construct as needed.

Science is founded on sorites. Through this type of deduction, anyone may generate a lot of information from a few remembered generalizations. Reliance on sorites permits an economy of experimentation. Quite possibly, no one has ever done an experiment to see if ravens need oxygen. Experiments *have* shown that diverse species of animals need oxygen, and had there been any reason to believe that ravens might be anaerobic creatures, that contingency would have been tested. As it is, we rely on the sorites above.

Scientists look for "all X's are Y's" generalizations because they lend themselves to quick deductions. The notion of a controlled experiment (where causes are isolated and identified with effects) presupposes that the important facts of the world are of this type. However, it does not follow that *all* truth can be formulated so simply. As we carve out a part of the truth, it is well to reflect that our slice of reality may not have the same shape as the whole.

Complexity

Holmes's complaint about the UND riddle in the previous chapter —that it is impractical to solve it "logically"—typifies the opposite type of logical process. There the step-by-step deduction of a sorites does not apply.

The UND riddle is apropos of a branch of mathematical logic known as *complexity theory*. Complexity theory studies how difficult problems are in an objective, abstract sense. It was founded on the experience of computer programmers, who discovered that some types of problems are much more difficult than others to solve by computer.

Complexity theory would be less useful if it applied only to computers. It applies no less to humans solving problems. A human must solve a problem by some method, and these methods (rather than hardware) are the concerns of complexity theory.

It may appear futile to look for an objective measure of how difficult a problem is. Most of the problems that arise in the real world are easy for some people and hard for others. Solution of many problems depends on making various mental connections between the problem and certain other facts. You either make these connections or you don't.

In a sense, riddles requiring a specific mental connection (such as Watson's land-division problem) are the most difficult sort of logic puzzles, for it is all but impossible to say *how* to solve them. In another sense, they are the easiest. If and when you make the connection, there is nothing to it.

Complexity theory is mostly concerned with problems that are difficult even when a methodical means of solution exists. There are inherently difficult problems that cannot be solved by the human mind or by the computers of a science-fictional distant future. Yet these problems are solvable, not paradoxes or "trick questions" without a solution.

A central notion of complexity theory is the *algorithm*. An algorithm is an exact, "mechanical" procedure for doing something. It is a set of directions so complete that no insight, intuition, or imagination is required. Any computer program is an algorithm. So is a recipe for vegetable soup, the directions for assembling a bicycle, and the rules to many simple games. The rules of arithmetic taught in elementary school are an algorithm too. You know that when you add two numbers, no matter how large, the rules will always produce a correct solution. If you get a wrong answer, you know you must have applied the rules incorrectly. No one doubts the algorithm itself.

An algorithm must be exact. "If you get lost in the woods, keep your cool, use common sense, and just play it by ear" is advice, but not an algorithm. The Boy Scouts' prescription—

If you get lost in the woods, walk downhill until you come to a stream. Then walk downstream; you'll eventually come to a town.

—is an algorithm.

It's tough to come up with effective algorithms. Unforeseen situations arise. It isn't hard to think of cases where the Boy Scouts' algorithm would fail. You could be in a desert basin where your walk downhill would lead to a dry lake bed, not a stream. In some remote parts of the world, there are streams that lead to a lake or the ocean without ever nearing human habitation. Worse yet, the instructions are silent on what to do if you find yourself on a plane so flat there is no obvious "downhill." An ideal algorithm would work no matter what the circumstances.

We do not always use algorithms. There are cooks who follow recipes, and there are cooks who improvise so freely that they claim to be unable to describe how a dish is made. Neither approach is right or wrong. Only the algorithmic approach lends itself to analysis, though.

Liars and Truth Tellers

Logic puzzles are a microcosm of the deductive reasoning we use to understand the world. Let's see how a logic problem can be solved methodically. One of the oldest genres of logic puzzles concerns the inhabitants of a distant island, some always telling the truth and some always lying. Members of the Truth Teller tribe always tell the truth. The Liars always lie. You must understand that there's no subtlety to the Liars: They don't try to conceal a lie by sometimes telling the truth. *Every* statement they make is the exact opposite of the truth. No characteristic dress or other clues allow outsiders to tell to which tribe a person belongs. Perhaps the most repeated Liars and Truth Tellers problem was devised by Nelson Goodman of grue-bleen fame and published (uncredited) in a Boston *Post* puzzle feature in 1931. Slightly modified, it goes like this:

On the island of Liars and Truth Tellers, you meet three people named Alice, Ben, and Charlie.

You ask Alice whether she is a Liar or a Truth Teller. She answers in the local dialect, which you do not understand.

Then you ask Ben what Alice said. Ben, who speaks English, says, "She said she's a Liar." You then ask Ben about Charlie. "Charlie is a Liar too," Ben insists.

Finally, Charlie adds, "Alice is a Truth Teller."

Can you figure out to which tribes the three belong?

Who Is Lying?

As in a syllogism, the basic logic of a Liars and Truth Tellers problem transcends the subject matter. Had the story began with the protagonist bailing out of a plane and landing on the island, it would make no difference. Had different names been chosen for the trio, it would make no difference beyond the fact that those names would appear in the solution. The essential problem is one of logical relationships, and these are all that really count.

You are interested in one thing only—determining the tribes of the three natives. When solving arithmetic problems, we often write something like $x = 12 + 5y$. There x and y are variables, unknown quantities that may have any of a range of possible values. Solving the problem is a matter of deciding what specific values x and y must have. A logic problem can be treated the same way. There are three unknowns in this logic puzzle: whether Alice is a Truth Teller, whether Ben is a Truth Teller, and whether Charlie is a Truth Teller.

You could just as well say the unknowns are whether Alice, Ben, and Charlie are Liars. It makes no difference, but let's give them the benefit of the doubt and work with the Truth Tellers formulation. Then we have three simple propositions that may be true or false:

> Alice is a Truth Teller.
> Ben is a Truth Teller.
> Charlie is a Truth Teller.

These statements are as fundamental as any can be about the situation. They are atoms of the situation's logic; no simpler statements exist. Since these sentences are not things we know for a fact, but only dummy statements that may be true or false, the role these sentences play is much like that of the variables of algebra. The "values" these sentences may have are, of course, *true* or *false*. In logician's jargon, such statements are called *Boolean variables,* after British logician George Boole (1815–64).

The problem's first question is directed to Alice. Since her reply is unintelligible to us, we can deduce nothing from it.

The first real information comes from Ben. He says that Alice said she's a Liar. As you probably gather, you can't take this at face value. Ben may be lying about what Alice said, and Alice may be lying about herself. Ben's statement is possible only under certain tribe assignments—under certain suppositions about who's a Liar and who's a Truth Teller.

Let's see. Alice and Ben can't *both* be Truth Tellers. If they were, Alice would have honestly said she was a Truth Teller, and Ben would have honestly translated the statement. Since Ben said Alice said she is a Liar, we know both aren't Truth Tellers.

Can Alice and Ben both be Liars? Yes. Alice, asked if she was a Liar, would say she was *not.* Then knee-jerk liar Ben would negate that for a double negative. Ben would say that Alice had said she *was* a Liar. That's just what he did say.

In fact, no one ever says, "I am a Liar." A Truth Teller would not tell that lie, and a Liar would not tell that truth. When asked straight out, everyone insists that he is a Truth Teller (as in real life).

Ben's statement that Alice said she was a Liar is a dead giveaway. No matter what Alice is, she must have said she is a Truth Teller. Ben is a Liar for saying otherwise.

(What if Alice didn't understand the question at all? Then she probably would have said, "I don't understand English," or—if a Liar!—"I do understand English." Ben would report one of these responses, the wrong one if he in turn is a Liar. Because the Liar tribe is so unimaginative, we know from Ben's actual answer that Alice must have understood the question and responded with a statement about her tribe.)

Since Ben is a Liar, his second statement ("Charlie is a Liar") must also be false. Charlie must be a Truth Teller.

That leaves Charlie's statement. He says that Alice is a Truth Teller. We already know that Charlie is a Truth Teller, so this must be the case. The solution is that Alice is a Truth Teller, Ben is a Liar, and Charlie is a Truth Teller.

Is there a method to this solution? Well, sort of. The realization that no one says he's a Liar helped. That revealed that Ben was a Liar, and then things fell into place.

But this method, if you can call it that, doesn't apply to any and all Liars and Truth Tellers problems. Take this simple but novel Liars and Truth Tellers problem of Raymond Smullyan's: A person of unknown tribe says, "I am a Liar, or 2 + 2 = 5." What is the person's tribe?

Now, the person is not saying he is a Liar. He is linking two statements with "or," which means that at least one of the statements is true—*if* the speaker is telling the truth.

There are two possible hypotheses about the speaker: that he is a Truth Teller and that he is a Liar. If the speaker is a Truth Teller, what he says is true. We can depend on the statement that the speaker is a Liar or 2 + 2 = 5.

This *can't* be the case. At least one of the two statements joined by "or" must hold for the compound statement to be true. There's no way that 2 + 2 = 5 can be true, so the part about the speaker being a Liar would have to be true. That contradicts the assumption that the speaker is telling the truth.

All right, try the other hypothesis. Assume the speaker is a Liar. Then it is *not* the case that the speaker is a Liar or 2 + 2 = 5. For an *or* statement to be false, both of the component statements must be false. If just one were true, the *or* statement would be true. Therefore, saying "A or B is false" is the same as saying "both A *and* B are false." If the speaker is a Liar, both these statements must be false: "the speaker is a Liar" and "$2 + 2 = 5$." Again we have a contradiction. If the speaker is Truth Teller, he must be Liar, and if he's a Liar, he must be a Truth Teller.

In fact, Smullyan's puzzle is a cleverly disguised form of the liar paradox. The "solution" is that no solution is possible. (Or as Smullyan put it, the only possible conclusion is that the problem's author is not a Truth Teller.)

There is a method that works for any Liars and Truth Tellers problem, even those that turn out to be insoluble like Smullyan's. For each of the islanders mentioned, there are two possible tribes: Truth Teller and Liar. Call any guess about each of the islanders' tribes (like "Alice is a Liar, Ben is a Liar, and Charlie is a Truth Teller") a "complete hypothesis." For any problem, there is a fixed number of complete hypotheses about the islanders ($2 \times 2 \times 2$ or 8 in Goodman's problem). All you need do is run through the hypotheses and see which are allowed by the statement of the puzzle.

In each case, you look for a contradiction, a reductio ad absurdum. For instance, in Goodman's problem the assumption that all three are truthful leads to the conclusion that Ben would say something other than what he did say. That's wrong, so cross that hypothesis off the list. After running through the eight possibilities, you would find that only one does not lead to contradiction: Truth Teller/Liar/Truth Teller for Alice, Ben, and Charlie, respectively. The problem is solved by the process of elimination.

"How often have I said to you that when you have eliminated the impossible, whatever remains, however improbable, must be the truth?" asked Sherlock Holmes in *The Sign of Four*. The process of elimination will solve many types of problems. But it is not always practical.

The trouble is, the process of elimination is slow. It's slow because the number of hypotheses is often overwhelming.

A Boolean variable can be only true or false. That's two possibilities per unknown. Each unknown doubles the total number of complete hypotheses. In a problem with three Boolean unknowns, the number of possible hypotheses is 2^3 or 8. In general, when there are n true-or-false unknowns, there are 2^n possible complete hypotheses. For a Liars and Truth Tellers with a couple dozen islanders, the number of hypotheses would number in the millions.

SATISFIABILITY

We arrive now at the heart of deductive reasoning. The anecdotal structure of logic problems—what the problems seem to be "about" —is irrelevant to solution. Abstract away the window dressing, and what remains behind?

It is SATISFIABILITY. To complexity theory, this is the fundamental, irreducible kernel of logic. It is a "skeleton" that exists inside every problem of deduction.

We appreciate that the problem of adding 273 apples to 459 apples is essentially the same as the problem of adding 273 oranges to 459 oranges or that of adding 273 croquet mallets to 459 croquet mallets. The realization that all such problems are fundamentally the same is the basis of arithmetic.

Complexity theory was founded on the realization that many more complex problems are really the same. Arithmetic arose in the accounting problems of the ancients. Adding and subtracting bushels of wheat, someone realized, wasn't any different from adding and subtracting mules or gold coins. Complexity theory was motivated by the problems confronting computer programmers in the 1960s and 1970s. These programmers discovered that many seemingly different problems were equivalent.

By convention, SATISFIABILITY is phrased as a yes-or-no question: Given a set of premises, are they compatible? Or: Do they describe a possible world? Or: Do they contain an irresolvable paradox?

A complete SATISFIABILITY problem includes a set of Boolean variables—fundamental statements whose truth or falsehood is initially unknown—and a set of logical statements about the Boolean variables. These statements may use the standard logical relations such as "or," "and," "not," and "if . . . then."

Often, each statement describes a single, ambiguous observation. Take Goodman's Liars and Truth Tellers puzzle. Symbolize the three Boolean variables by the individuals' names. The problem in essence says,

Boolean Variables: Alice (meaning Alice is a Truth Teller)
Ben (meaning Ben is a Truth Teller)
Charlie (meaning Charlie is a Truth Teller)
Statements: 1. *if* (Ben *and* Alice) *then not* Alice

2. *if* Ben *then not* Charlie
3. *if* Charlie *then* Alice

The first and trickiest statement corresponds to Ben's assertion that Alice said she was a Liar. Alice is not a Truth Teller, provided that both Ben and Alice are Truth Tellers so that the statement can be trusted.

The Pork-Chop Problem

SATISFIABILITY problems can be difficult indeed. Lewis Carroll constructed ponderous logic puzzles in which the solver is required to deduce a single valid conclusion from a dozen or more nonsensical premises. Several are included in his unfinished textbook, *Symbolic Logic*. The problems are absurd travesties of scientific or mathematical reasoning, and amazingly difficult as well. The more difficult problems are beyond the limits of most people's patience (though they have been solved by computer). The most difficult, discovered in his notes and not published until 1977, contains fifty premises.

One that has been extensively analyzed by hand and by computer is the notorious "pork-chop problem." The puzzle is to derive the "complete conclusion," a hypothesis that is compatible with and demanded by all the other statements.

THE PORK-CHOP PROBLEM

(1) A logician, who eats pork-chops for supper, will probably lose money;[1]

(2) A gambler, whose appetite is not ravenous, will probably lose money;

(3) A man who is depressed, having lost money and being likely to lose more, always rises at 5 A.M.;

(4) A man, who neither gambles nor eats pork-chops for supper, is sure to have a ravenous appetite;

(5) A lively man, who goes to bed before 4 A.M., had better take to cab-driving;

---

[1] For us today, the problem is further obscured by ambiguous wording and punctuation. Carroll set off restrictive clauses with commas, against modern usage. His notes indicate that this first premise was to be understood as *"all logicians who eat pork chops will probably lose money."* When "though" occurs (as in premise 8), it should be interpreted as a logical "and."

(6) A man with a ravenous appetite, who has not lost money and does not rise at 5 A.M., always eats pork-chops for supper;

(7) A logician, who is in danger of losing money, had better take to cab-driving;

(8) An earnest gambler, who is depressed though he has not lost money, is in no danger of losing any;

(9) A man, who does not gamble, and whose appetite is not ravenous, is always lively;

(10) A lively logician, who is really in earnest, is in no danger of losing money;

(11) A man with a ravenous appetite has no need to take to cab-driving, if he is really in earnest;

(12) A gambler, who is depressed though in no danger of losing money, sits up till 4 A.M.;

(13) A man, who has lost money and does not eat pork-chops for supper, had better take to cab-driving, unless he gets up at 5 A.M.;

(14) A gambler, who goes to bed before 4 A.M., need not take to cab-driving, unless he has a ravenous appetite;

(15) A man with a ravenous appetite, who is depressed though in no danger of losing money, is a gambler.

We are used to thinking of logic as something that comes naturally. You expect to hit on a logic problem's answer without really thinking about how you arrive at it. In Carroll's problems, the statements are much too numerous and illogical to grasp at once. You have to resort to algorithms such as the tree diagrams and registers that Carroll described (or to computer programs).

The pork-chop problem has 11 Boolean variables (being earnest; eating pork chops; being a gambler; getting up at 5 A.M.; having lost money; having a ravenous appetite; being likely to lose money; being lively; being a logician; being a man who had better take to cab-driving; sitting up till 4 A.M.). There are 2^{11} or 2,048 distinct hypotheses about an arbitrary individual.

In asking for a conclusion, the pork-chop problem resembles a scientific investigation. It seems rather different from SATISFIABILITY, a yes-or-no question. That SATISFIABILITY is a yes-or-no question does not limit its generality, though. As in the game "20 Questions," any information can be imparted by a series of yes-or-no answers. A logic problem posing an arbitrary question can be rephrased as one or more yes-or-no problems.

Let's say you want to test the conclusion: "A man who eats pork chops is lively." Step one is to take the original 15 premises as a SATISFIABILITY problem. Are they mutually compatible? The answer should be yes. Otherwise it's not a fair problem. Then add the proposed conclusion as a 16th statement. Ask if the amended list of statements is still compatible (a second SATISFIABILITY problem). If so, the new statement is at least permitted by the original premises.

That doesn't necessarily mean it is a valid conclusion. You could test "The moon is made of green cheese" as the 16th statement, and the set would of course be satisfiable. Since it says nothing about logicians, gamblers, or the other elements of Carroll's nonsense, it cannot possibly introduce a contradiction.

A third SATISFIABILITY problem is needed to make sure that a hypothesis is *demanded* by the original premises. Replace the hypothesis with its negation, its logical opposite: "Not all men who eat pork chops are lively." With this negation as the new 16th statement, check to see if the set is compatible.

If a hypothesis *or* its exact negation can be added to the premises without contradiction, then clearly the hypothesis is irrelevant. Both "The moon is made of green cheese" and "The moon is not made of green cheese" are compatible with the pork-chop problem, so neither is a valid deduction.

If instead a hypothesis is compatible with the premises and its negation is not, then it is a proper conclusion. (If a hypothesis is not compatible but its negation is, then the negation is a valid conclusion.)[2]

Like all general problems, SATISFIABILITY is easy *sometimes*. It can be easy even when the number of Boolean variables and clauses is huge.

It is not always necessary to check every possibility, or even most of them. Often, many statements can be linked into sorites. If this is so, then that is so, and if that is so, then this is so . . . Such deduction has great power to "make sense" of a large number of statements.

Each link of a sorites can be expressed as an *if . . . then* statement concerning two Boolean unknowns. When the statements of a SATISFIABILITY problem mention just two Boolean variables each, the problem is easy. There are efficient methods for solving

<hr>

[2] In case you're wondering, the answer to the pork-chop problem is: "An earnest logician always gets up at 5 A.M. and sits up till 4 A.M."

the problem that are much faster than running through every possible hypothesis to find the right one.

Not all logic problems are so easy. When statements connect three or more Boolean unknowns, there is no general solution significantly faster than the process of elimination. Carroll's pork-chop problem is notably hard because the premises link three or four Boolean variables (such as logicians, pork chop eaters, and money losers).

The Elevator Problem

The increase in difficulty once statements concern three unknowns is evident in the "elevator problem":

Six people are in an elevator. At least three of them are mutual acquaintances, *or* at least three are complete strangers to one another. Can you prove that this is *always* so?

This is true, but it is difficult to prove it "logically." Common-sense reasoning about strangers and acquaintances doesn't get you anywhere. You cannot deduce that A knows B from B knows C, since the problem says nothing about pairs of persons, only triplets.

The elevator problem exists in many versions. The six people can be the ill-chosen guests at a dinner party, some of whom aren't speaking to one another because of past feuds. If no three guests are all on speaking terms, prove there is a trio who aren't speaking to one another at all. A bawdy version of the problem claims that, of any six residents of a college dormitory, at least three have slept with each other, or at least three have never slept with each other.

The elevator problem illustrates a branch of mathematics called *graph theory*. Graph theory enters (often unrecognized) into many problems, both recreational and practical. One of the best-known is the "gas, water, and electricity problem" popularized by Henry Ernest Dudeney, author of puzzles and riddles for newspapers and magazines around the turn of the century. The answer to the original version of the problem is that there is no answer. It is impossible to connect three dots to three other dots in the plane without at least one connection crossing another. When puzzles like this were in their heyday, no one was overly concerned that unsuspecting readers might spend hours or days on an unsolvable puzzle. Of course, Watson and Holmes's clever solution in the previous chapter is beside the point!

Graph theory is not about the sort of graphs that show stock-market averages or annual rainfall. A graph in graph theory is a network of points connected by lines, like the route maps of airlines you see in airports. Whether the lines are straight or wiggly makes no difference, nor does the relative position of the points. Only the topological properties of the network count—*which* points are connected by lines. All this is true enough, but it says nothing about why such things are important or useful. In a broader sense, graph theory is the study of relationships or connections between elements.

The elevator problem is easily translated into a graph. Represent the six people as dots (see diagram). Between every two dots you can draw a line representing a relationship. Let a black line mean the pair are acquaintances, and a gray line mean that they are strangers. A trio of mutual acquaintances appears as a black triangle; a trio of mutual strangers as a gray triangle. Is it possible to draw in lines between all the dots so that no all-black or all-gray triangles appear?

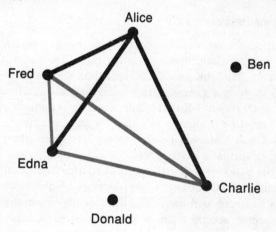

The proof is easy to follow. Start with Alice. We will draw five lines from Alice, representing whether she knows or doesn't know the other five persons in the elevator. No matter what, at least three of the lines must be the same color. That's because there are five lines and two colors. The evenest split gives you three lines of one color and two of another. Otherwise, there will be four or even five lines of one color.

We don't know whether Alice will know at least three persons (black lines), or not know at least three persons (gray lines). Take the first eventuality. Let's say that three black lines connect Alice with Charlie, Edna, and Fred. How do we color the lines connecting this latter trio?

If any of these three lines are colored black, then that produces an all-black triangle: a set of three people who are mutual acquaintances. The only way to avoid an all-black triangle is to color the lines between Charlie, Edna, and Fred gray. That produces an all-gray triangle, which, of course, is a set of three persons who are strangers. Either way, the trio of acquaintances or strangers is inevitable.

If instead Alice *doesn't* know three or more of the others, similar reasoning leads to an identical conclusion. An all-black or all-gray triangle must exist.

This is not the only case where a logic problem is equivalent to a geometric one. Complexity theory recognizes that many different types of problems are procedurally identical.

Science and Puzzles

Metaphors for the scientific process are many: a riddle, a cipher, a jigsaw puzzle. Confirmation is often more like the solution of a logic puzzle than the models of induction discussed in previous chapters. A simple generalization is confirmed or refuted by any relevant observation. Most scientific theories are much more complex and must be evaluated in light of a large number of observations. You may not even be able to say if a given observation, in isolation, confirms or disconfirms.

Take the hypothesis that the earth is round. Confirming this was not a matter of assembling a large number of observations of the round earth (from astronauts?) in the absence of counterexamples. Rather, people accepted the earth's roundness because it related and made sense of experiences that previously seemed meaningless. To the ancients, few items of trivia could have seemed as unconnected as these: the midnight sun of the Far North; the round umbra of the lunar eclipse; the way ships seem to sink beneath the waves as they recede from port. Now all are seen as logical consequences of the roundness of the earth. It is because the round-earth theory explains so many unrelated observations that it is so compelling. Only by an incredible coincidence could all these observations agree so neatly with a round earth if the earth was not in fact round.

This more subtle type of confirmation combines deduction with induction. A hypothesis has logical consequences that must first account for past observations and then make new predictions. True predictions confirm the hypothesis. The interplay of induction and deduction is the source of paradoxes even more baffling than those discussed so far.

6

BELIEF

The Unexpected Hanging

A PRISONER APPEARS before a hanging judge for sentencing. "I am not allowed to dispense cruel or unusual punishment," the judge begins, rather inauspiciously. "The harshest punishment I am permitted to recommend is hanging by the neck until dead. The gallows it must be then. Beyond that, my only freedom is in setting a date for your hanging. I am of two minds on that.

"My impulse is to order an immediate execution and be done with it. On the other hand, that might be an undeserved kindness. You would have no time to contemplate your impending doom. I choose instead this compromise: I sentence you to hang at sunrise on one of the seven days of next week. I further instruct your executioner to make certain that you have no way of knowing in advance on which day you will be hanged. Every night, you will go to sleep wondering if the gallows awaits the next morning, and when you do walk the last mile, it will come completely as a surprise."

The prisoner was taken aback to find his lawyer smiling at this incredibly cruel sentence. When they got out of the courtroom, the lawyer said, "They can't hang you." He explained: "You are supposed to be hanged at sunrise on one of the seven days of next week. Well, they can't hang you on Saturday. It's the last day of the week, and if you aren't hanged by Friday morning, then you can know with utter certainty that the day of execution is Saturday. *That* would violate the judge's plan of not letting you know the day in advance."

To this the prisoner agreed. The lawyer continued: "Therefore, the last day they can hang you is actually Friday. Fine. But look— they can't hang you Friday either. Granting that Saturday is *really* out of the question, Friday is the last day they can hang you. If you make it to breakfast Thursday morning, you will know for a fact that you are to die Friday. And *that* is against the judge's orders. Don't you see? The same logic rules out Thursday, Wednesday, and every other day. The judge has outsmarted himself. The sentence is impossible to carry out."

The prisoner rejoiced until Tuesday, when he was awoken from a deep sleep and sent to the gallows—quite unexpectedly.

Pop Quizzes and Hidden Eggs

The paradox of the "unexpected hanging" packs a double whammy. You think the paradox is that the seemingly plausible sentence can't be carried out—and then it is. Philosopher Michael Scriven wrote of it: "I think this flavour of logic refuted by the world makes the paradox rather fascinating. The logician goes pathetically through the motions that have always worked the spell before, but somehow the monster, Reality, has missed the point and advances still."

The paradox has the fairly unusual distinction of being inspired by a real event. It dates to a wartime (1943 or 1944) radio announcement of the Swedish Broadcasting Company:

> A civil defense exercise will be held this week. In order to make sure that the civil defense units are properly prepared, no one will know in advance on what day this exercise will take place.

Swedish mathematician Lennart Ekbom recognized the subtle contradiction and mentioned it to his class at Ostermalms College.

From there it soon spread around the world. It has gone through several anecdotal incarnations. Others versions talk of a "class A blackout," a surprise military exercise to take place the following week, or a teacher's anticipated "pop quiz."

Shades of this paradox occur in many situations where one person's knowledge is incomplete. E. V. Milner noticed an analogy in the New Testament parable of Dives and Lazarus. Dives, a rich man, goes to hell while poor Lazarus, who has suffered all his life, goes to heaven. Dives pleads with Abraham for mercy but is told no, the injustices suffered in life are exactly compensated for in the afterlife. Those who were fortunate in life must suffer. Milner's paradox of Dives and Lazarus explores this somewhat ironic concept of otherworldly justice:

> . . . suppose, in fact, that some means were found to convince the living, whether rich men or beggars, that "justice would be done" in a future life, then, it seems to me, an interesting paradox would emerge. For if I *knew* that the unhappiness which I suffer in this world would be recompensed by eternal bliss in the next world, then I should be happy in *this* world. But being happy in this world I should fail to qualify, so to speak, for happiness in the next world. Therefore, if there were such a recompense awaiting me, its existence would seem to entail that I should at least not be wholly convinced of its existence. Put epigrammatically, it would appear that the proposition "Justice will be done" can only be true for one who believes it to be false. For one who believes it to be true, justice is being done already.

One minor weakness in the prisoner paradox is the possibility that the prisoner may not be hanged at all. The inevitability of the execution is essential to the prisoner's deductions. To avoid this weakness, Michael Scriven's 1951 analysis in the British journal *Mind* restated the paradox as an experiment with an egg. In front of you is a row of ten boxes marked No. 1 through No. 10. While your back is turned, a friend hides an egg in one of them. There is no doubt of that; the egg is there somewhere. The hider says, "Open the boxes in order. I guarantee that you will find an unexpected egg in one of them." Of course she can't hide the egg in box No. 10, for after opening box No. 9, you would know the egg's location. Deductions and counterdeductions follow, as does the surprise at finding the egg in, say, box No. 6.

Hollis's Paradox

There is no limit on how far the prisoner's logical daisy chain can reach. Look at this recent variation, "Hollis's paradox" (for Martin Hollis):

Two persons on a train, A and B, each think of a number and whisper it to fellow rider C. C gets up and announces, "This is my stop. You have each thought of a different positive integer. Neither of you can deduce whose number is bigger." C then gets off the train.

A and B continue their travel in silence. A, whose number was 157, thinks, "Obviously B didn't choose 1. If he did, he'd know that my number was the bigger, just from C's statement that we chose different numbers. Just as obviously, B knows I didn't choose 1. Yeah, 1 is completely out, for both of us. The smallest number that is even a possibility is 2. But if B had 2, he'd know that I didn't have it either. So 2's out . . ."

If A's train ride is long enough, he can rule out *every* number.

A Minimal Paradox

When in doubt, simplify. The seven days or ten boxes (or aleph-null integers!) are excess baggage. The paradox would "work" with six days/boxes, or five, or four. How far can it be trimmed? To two days? One day?

Let's try it with one day. The judge sentences the prisoner to be executed on Saturday (the prisoner, of course, hears this). The prisoner is also not supposed to know the day beforehand. Of course, he does know it. The only way the executioner could surprise him would be not to hang him at all. That is ruled out from the start. Consequently there can be no surprise and no paradox. The judge has asked for something that cannot be. Saying "You will die on Saturday, and it will be a surprise" is like saying "You will die on Saturday, and $2 + 2 = 5$." The second part of the sentence is wrong, that's all.

Go a little lighter on the pruning shears. Take the two-day case. The judge sentences the prisoner to die the following weekend, but the prisoner must not be able to deduce which day, Saturday or Sunday, beforehand. Does the paradox still exist?

It is again agreed that the prisoner will be executed on one of the two days no matter what. Sunrise comes Saturday with no hangman. Then at breakfast Saturday the prisoner knows with certainty that he will be executed Sunday.

That, however, means that the sentence will *not* be carried out as specified: There will be no surprise. Conclusion: It is not possible to carry out the sentence by hanging the prisoner on Sunday.

Can the surprise requirement be met on Saturday? Well, it depends on whether the prisoner expects a Saturday execution. There are two possibilities. Either the prisoner expects the hangman on Saturday, or he doesn't.

The prisoner might figure, "Well, I'm a goner all right," and leave it at that. He may not have any opinion about which day it will be. In that case, the executioner need only hang him on Saturday to satisfy the judge. (Sunday is still out. Even the most stoic prisoner will realize that he will die Sunday if not hanged Saturday.)

The paradox's hook lies in the alternative, that the prisoner *does* analyze his plight and expects the hangman on Saturday. Then the executioner will have failed to meet the surprise provision.

Forget that this a logic paradox for a moment. What would you do if you were the executioner? You have to execute the prisoner on Saturday or Sunday, and you have to follow the judge's orders if at all possible.

Apparently, a reasonable executioner trying his level best to comply with the orders would almost have to choose Saturday for the hanging. There's no chance of a Sunday hanging being unanticipated. At least with Saturday the executioner can hope that the prisoner didn't give the matter much thought.

So the executioner takes the prisoner to the gallows at sunrise Saturday. As is customary, the prisoner is allowed to say his last words. He turns to the judge and says, "Your executioner failed! I expected to be hanged today. Only by hanging me today would there be any chance of my not anticipating it. But I did!"

Prisoner and executioner are locked in a battle of wits: Each can anticipate any reason the other might think up for choosing any specific day. The paradox can be short-circuited with a "stupid" prisoner, who doesn't contemplate his fate or try to second-guess. But when prisoner and executioner are the perfect logicians of a logic puzzle, the paradox is profound indeed.

A Time-Travel Paradox

Scottish mathematician Thomas H. O'Beirne pointed out that it is possible for one person to make a prediction about future events which is true but which others do not know to be true until afterward. The judge is correct in saying that the prisoner will be surprised, even though the prisoner doesn't know it (yet).

This can be made clearer by reformulating the paradox thus: The judge sentences the prisoner to die sometime the following week (leaving the date to the executioner). Then the judge hops in a time machine and sets the dial for a week or more hence. Arriving in the near future, the judge steps out, gets a newspaper, and reads that the prisoner was hanged the Tuesday after sentencing. In the prisoner's final interview, he said he was surprised at the date; he thought they'd let him live until the end of the week. A cruel idea pops into the judge's head. "Suppose I go back to the day of sentencing and tell the prisoner that he won't be able to guess the day of his execution," he thinks. "That will be a correct statement because, here in the future, I know he was surprised. And just telling him that will drive him crazy!"

The judge gets in the time machine again and returns to the day of sentencing. He gets out and says to the prisoner (as in the original paradox), "You will be hanged next week but you will not be able to guess the day of execution beforehand." The prisoner concludes he can't be hanged, and he is wrong; the judge is right.

Anything wrong here? Well, the judge really saw the consequences of his *original* sentence (which did not say anything about the date being unexpected). Telling the prisoner that he will be surprised changes things—possibly insignificantly, possibly a lot. The prisoner's surprise is no longer certain.

While in the future, the judge might also learn that a surprise party he threw for his sister on her birthday was unexpected. If he went back to the previous week and told his sister that, then obviously she would not be surprised. Imparting *some* valid information about the future can invalidate the information.

If the judge is free to use the time machine as he likes, this may not be much of a limitation. After telling the prisoner that he will be surprised, the judge can zip back into the future to make sure that the prediction still holds. If it does, fine. If not, he can come back and modify the sentence until the prediction and the reality coincide. The result would be a prediction which is true but which the prisoner cannot know to be true until after the fact.

Though it seems quite different, "Berry's paradox" (after librarian G. G. Berry, who described it to Bertrand Russell) has something of the same flavor. Think of the least integer not namable in fewer than nineteen syllables. Certainly some integer qualifies as just that. But the phrase "the least integer not namable in fewer than nineteen syllables" is a description of a certain number, and that description has eighteen syllables. Ergo, the least integer not namable in fewer than nineteen syllables can in fact be named in eighteen syllables!

Berry's paradox defies facile resolution. Once you're ready to write it off to ambiguous wording, bring in that stock character of paradoxes, the omniscient being. The being seemingly could be aware of every possible description of every number or sentence. To this being, one number *is* the least integer not namable in fewer than nineteen syllables! The being, like the judge, seems to know something forbidden to us.

All this perhaps shows that the judge can know what he is supposed to know in the paradox. More interesting, though, are the deductions of the prisoner and the executioner. Who, if anyone, is right?

What Is Knowledge?

The paradox of the unexpected hanging poses the question: What is knowledge? The prisoner is caught up in a web of second-guessing, third-guessing, and nth-guessing. He *thinks* he knows he can't be hanged on Saturday. The executioner thinks he knows that the prisoner *can't* know the day of his execution. The paradox raises the twin specters of being right for the wrong reason or being wrong for the right reason. These same concerns frequently pop up in the philosophy of science. There (more than in criminal justice!) we often "know" things by chains of reasoning as convoluted as the prisoner's.

Like most common words, "know" has a lot of flexibility built into it. We all say things like "I know the Celtics are going to win the championship" when there is actually considerable doubt. In science, you usually want to know things with greater certainty than that.

For years philosophers defined knowledge by a set of three criteria called the "tripartite account." The idea was that these criteria would be met if, and only if, you knew something.

Let's take an example from the supposedly certain realm of mathematics. Suppose you know that 4,294,967,297 is a prime number (a number that cannot be divided evenly by any whole number other than 1 and itself). Then these three conditions must hold:

First, you *believe* that 4,294,967,297 is prime. If you don't even believe it, you can't know it. We wouldn't say that a flat-earther knows the earth is round.

Second, your belief that 4,294,967,297 is prime is *justified.* You must have good reason for believing it. You can't believe it just because of a mistake in arithmetic. Nor can you believe it on the basis of a hunch, or reading tea leaves, or a fit of temporary insanity.

Third, 4,294,967,297 really *is* prime. Obviously, if the statement is wrong, you don't know it as fact.

The first time you hear of it, the tripartite account sounds almost too trite to be of any interest. But knowledge is a more complex business than it appears. The second criterion is most troublesome of the three. Why require "justified" belief at all? It might seem that it is enough to believe something and have it be true.

A two-criteria definition would include anyone who "lucked out" and was right about something for bad reasons. After Kennedy's assassination (1963) and the Reagan assassination attempt (1981), several psychics came forward with claims that they had predicted the events. At least some had made predictions about the Presidents being in peril on the approximate dates, and these predictions had been published or appeared in press releases before the event. The same psychics also made scores of other predictions that did not come true. Washington psychic Jeane Dixon makes so many predictions each year that she cannot help being right many times. If this is "knowledge," it is not a very useful kind.

It is not always easy to say what constitutes "good reason" for believing something. In 1640 French mathematician Pierre de Fermat felt he had reason to believe that 4,294,967,297 is prime. He noticed that you can produce prime numbers with the formula

$$2^{2^n} + 1.$$

Fermat's formula has a multi-level exponent. A regular exponential expression, like 2^3, means the lower number (2) multiplied by itself the number of times indicated by the little superscript number (3). 2^3 is $2 \times 2 \times 2$, or 8. In Fermat's formula, you choose any number for n, evaluate the topmost expression (2^n) and multiply the bottom 2 by itself *that* many times. Then add 1.

For instance, $2^{2^1} + 1$ is 5, and 5 is prime. $2^{2^2} + 1$ is 17, $2^{2^3} + 1$ is 257, $2^{2^4} + 1$ is 65,537, and all are primes. Fermat suspected that 4,294,967,297 ($2^{2^5} + 1$) and all the higher members of this series must be prime.

Lots of other people believed that too. There was empirical evidence and the backing of authority. But as you've probably guessed, 4,294,967,297 isn't prime at all. Swiss mathematician Leonhard Euler discovered that it is 641 times 6,700,417.[1]

Science and the Tripartite Account

Belief, justification, truth—the history of science includes examples of all permutations of these three conditions. Let's use T to indicate that a condition holds and F to indicate that it doesn't, and list the criteria in the order above.

TTT stands for a justified true belief or what is held to be true knowledge. In this category goes most scientific belief; that portion of it that is right, anyway.

FTT represents a justified truth that is disbelieved. There are many instances of this; one is creationism, the quasi-scientific cult that rejects evolution in the face of overwhelming evidence. The fuddy-duddyism of those who reject new discoveries (the French Academy's refusal to accept meteorites; physicist Herbert Dingle's crankish refutation of relativity) is FTT. So is the inertia that made physicist Max Planck complain (1949): "A new scientific truth does not triumph by convincing its opponents and making them see the light, but rather because its opponents eventually die, and a new generation grows up that is familiar with it."[2]

TFT is an unjustified true belief; being right for the wrong reasons, as in a psychic's lucky guess. Again, there are many examples. Democritus of the fifth century B.C. held the true belief that all matter is made of particles so tiny as to be invisible: atoms. Although Democritus' works are lost, it is unlikely he had anything we would consider valid evidence. His was a philosophical insight that turned out to be right. (The serendipity is less striking for the fact that the atoms of twentieth-century physics are not indivisible as Democritus thought.)

TTF is a justified belief that is *wrong*. It is interesting to reflect that many successive cosmological views have been in this category. The ancients had the justification of their own senses for believing that the sun moved around the earth. Although generations of schoolteachers have cited this as the epitome of wrongness, it took a

---

[1] There are a number of formulas that produce primes for a while and then fail. One of the best-known is $n^2 - 79n + 1601$, which works for all values of n up to 79, then produces a nonprime number at 80. Such are the dangers of inductive generalizations in mathematics.

[2] To which Alan L. MacKay replied: "How can we have any new ideas or fresh outlooks when ninety percent of all the scientists who have ever lived have not yet died?"

certain intellectual daring to postulate that the sun is a physical body that circles the world at a great distance and creates night and day. When Copernicus placed the sun at the center of the universe, he had justification, but was just as wrong. Since TTF beliefs are false, I cannot cite a generally accepted current belief as an example. It should come as no surprise if large parts of our present cosmology are wrong too.

The four remaining permutations are cases that fail in at least two criteria for knowledge. TFF is an unjustified belief that is indeed wrong: superstitions, old wives' tales. FTF is the peculiar case of a falsehood that is disbelieved in spite of justification. This would describe the people who are skeptical of the TTF cases above. The Catholic hierarchy *disbelieved* Copernicus' *justified* but ultimately *false* belief that the sun is the center of the universe.

FFT is a truth that is disbelieved because of lack of justification: the justifiable skepticism of someone who rejects something that nonetheless turns out to be true. The generations of philosophers who rejected Democritus' belief in atoms (since they had no reason for believing in them) is an example. At some point in every scientific revolution, justified conservatism (FFT) turns to mere reaction (FTT).

The final case, FFF, is an unjustifiable false belief that is rejected: The disbelief of perpetual-motion machine skeptics, or anyone's disbelief of nonsense propositions like "The moon is made of green cheese."

Buridan Sentences

Some beliefs fail to fall into any of these categories. "Buridan sentences," named after examples in fourteenth-century philosopher Jean Buridan's *Sophismata,* challenge any definition of knowledge. One goes:

> No one believes this sentence.

If this is true, no one believes it and therefore no one knows it. If it is false, at least one person believes it but no one (believer or nonbeliever) knows it because it *is* false. Consequently, no one can possibly know that the above sentence is true!

Would you believe this?

> You don't believe this sentence.

It would be silly to believe this sentence, since then you would be believing that you don't believe it. But if you don't believe it, then you have every reason to believe it because it is true . . . If the preceding sentence convinces you to believe it, that upsets the applecart again. Then it is absurd to believe it, all over again. Strangely enough, you can never arrive at a consistent position on this sentence. Yet, at any instant, an omniscient being aware of your every thought *can* say whether you believe it.

The opposite sentence ("You believe this") is the gist of Descartes's *Cogito ergo sum.* If only you believe the sentence, then it is true. If you disbelieve it, then it is false, and you have excellent ground for disbelieving it. No matter what your opinion of this sentence, you're correct.

Stranger yet is the "knower's paradox." It centers on this assertion (which is akin to the judge's statement in the unexpected hanging):

No one knows this sentence.

If *this* is true, no one knows it. If it is false, there is an immediate conflict: Someone knows it, but obviously no one can know a falsehood. Consequently, the sentence isn't false. It's an indubitably true fact that no one can ever know!

Gettier Counterexamples

Although the three conditions of the tripartite account have already led to paradox, they are not enough. They do not guarantee knowledge. There are ways you can have a justified true belief and *not* know what you believe. These ironic situations are known as Gettier counterexamples, after the American philosopher (Edmund Gettier) who discussed them in a 1963 paper.

As with an inductive generalization, a counterexample here is something that disproves a statement or line of argument. Gettier counterexamples are (usually) fictional situations that demonstrate that the three conventional criteria do not necessarily indicate knowledge. If the psychics mentioned above may be "right for the wrong reasons," then the essence of a Gettier counterexample is being "right for the right reasons, but the reasons don't apply." Errors of this type have intrigued philosophers (and storytellers) for a long time. Gettier counterexamples typically have an O. Henry flavor of farfetched coincidence.

Plato anticipated Gettier in one of his Socratic dialogues,

Theaetetus. There he discusses a lawyer so glib he can convince jurors of his client's innocence even if the client is guilty. Suppose the client is innocent. The jury believes the client is innocent and can cite the valid evidence they have just heard. But spellbound by gilded oratory, they would just as readily have believed a guilty client innocent. Plato contended that it is a false kind of knowledge they have. They really don't know the client is innocent.

One of Gettier's original illustrations was this: Smith and Jones are applying for a job with a company. Smith has just spoken to the president of the company and learned that Jones will get the job. Smith believes that Jones will get the job, and for good reason. Smith also believes that Jones has ten coins in his pocket. He just saw Jones empty his pocket looking for a quarter, and put ten coins back in the pocket. Smith has been watching him ever since and is sure he neither removed nor added any coins.

Smith muses idly to himself, "Well, it looks like the person who will get the job has ten coins in his pocket." He justifiably believes this, since it follows logically from the beliefs that Jones will get the job and Jones has ten coins in his pocket.

Gettier realized that these beliefs could be wrong, yet Smith could still be right. Suppose Smith gets the job (the company president changed his mind), and Jones actually has eleven coins in his pocket (one was stuck in the lining). Furthermore, it turns out that Smith has ten coins in his pocket. Then "the person who will get the job has ten coins in his pocket." It is ridiculous to say Smith knew this; it is sheer luck that it is true.

A Gettier counterexample need not be so contrived. Someone returns from lunch and asks you what time it is. You look at your watch and answer: 2:14. Your belief that it is 2:14 is certainly justified: It's an expensive watch that has always been reliable, and (something of a fanatic about the correct time) every night you set the watch by a government radio station that broadcasts the current time to the second. In fact, it *is* 2:14 P.M., but unknown to you, your watch stopped dead at 2:14 A.M. the previous night. By coincidence, you didn't look at your watch until that one minute out of every twelve hours that a broken watch is correct.

Another example: You go to the Louvre to see the *Mona Lisa.* You recognize the painting from a hundred photographs, and get goose bumps because you are in the same room with the *Mona Lisa.* You later find out that the museum staff, acting on a tip that someone would try to steal the painting, replaced it with a masterful reproduction the day you were at the Louvre. But you *were* in the same room with da Vinci's masterpiece because the real *Mona Lisa* was cleverly hidden behind a worthless painting nearby—the last place the thieves would look for it!

There have been Gettier counterexamples in the history of science. One is the alchemists' belief that metals can be transmuted into gold. This belief was founded on more than a mere hunch. As the first to systematize knowledge of substances, the alchemists were rightly impressed by how one substance may change into a profoundly different substance in chemical reactions. They further realized that the world is not infinitely diverse but is made of a relative few basic substances. If cinnabar can be turned into mercury, why not a base metal into gold? It seemed to be only a matter of hitting on the right combination of substances.

Even in retrospect, this was a plausible speculation. It just happened to be wrong. Brittle red cinnabar can be turned into silvery liquid mercury because it is a compound of mercury and sulfur (both elements). It *would* be possible to turn common substances into gold provided that gold is a compound of common elements, or else that some common substance is a compound of gold and something else. In fact, gold is an element and, unfortunately, no common substance is a compound of gold. Chemists can make gold from, say, gold chloride, but gold chloride is rarer than gold itself.

Despite this, it so happens that other elements can be transmuted into gold (or any element) in atomic reactions—of which the alchemists knew nothing. The alchemists had a justified true belief, yet it is surely wrong to say they "knew" other elements can be transmuted to gold.

One reaction to Gettier counterexamples is that they are really just unusual cases of being right for the wrong reasons. In each situation, the "justified" belief is not justified beyond all doubt. The likely is being confused with the certain.

Smith's conversation with the company president must not have been adequate grounds for believing that Jones would get the job. It was reason for assigning a high probability to Jones's getting the job, but not for believing it with certainty. Smith should have realized that executives are capable of reversing a decision and of purposely misleading a job candidate about his chances.

On the other hand, Gettier situations can be devised even for beliefs as certain as any belief about the external world can be. What are you most sure of right now, this very instant? You may be pretty sure that this book is in front of you right now. But you could be a disembodied brain in a vat. A laboratory janitor left a book in front of your vat while cleaning up, and by a wild coincidence, the book was *this* book.

The point is that if we demand that a "justified" belief be one of which we are certain, we short-circuit our attempt to define knowledge. Then one of the criteria for being certain of something would be to have reasons that make it certain. Even worse: *Nothing* in the external world is incontestably certain. If we have to be 100 percent certain of something to know it, then we don't know anything (not even the true things we justifiably believe).

A Fourth Condition

Much effort has gone into finding a "fourth condition" of knowledge. This would be an additional criterion that, combined with the first three, guarantees knowledge. It would have to eliminate all Gettier counterexamples, and not permit even more exotic counterexamples.

No one has yet come up with a fourth condition so obviously right that everyone has rushed to accept it. Of several attempts to frame a fourth condition, the most discussed holds that a justified true belief must also be *indefeasible*—it cannot be disqualified by extenuating circumstances.

Gettier's victims of false knowledge end up hitting themselves on the side of the head and saying, "If only I had known!" The victims could have avoided error if they had known—or just believed—certain information (that the painting was removed; that the watch had stopped; etc.). These disqualifying facts are called defeaters. Had Gettier's victims believed the defeaters, they would not have had justification for believing the paradoxically true statements.

It is 2:14 P.M.; you believe it is 2:14 P.M., having looked at your watch; you also believe that your watch stopped last night and hasn't run since. Then your belief that it is 2:14 is irrational. It is irrational because the defeater casts the original evidence of the time (your watch's hands indicating 2:14) in an entirely new light. The watch's hands are now irrelevant. The condition of indefeasibility requires that there *not* be an extenuating circumstance like this.

No one ever really knows when a belief is threatened by a defeater like this. The condition of indefeasibility may satisfy a theoretical need for a fourth condition, but it cannot help us avoid Gettier's false knowledge.

The Prisoner and Gettier

Now back to prisoner, judge, and executioner. You can argue from the tripartite account that the prisoner's "knowledge" is an illusion. W. V. O. Quine felt that *all* the prisoner's (or attorney's) deductions are wrong. Even the first deduction, that the prisoner cannot be hanged the last day, is invalid.

When the judge said the prisoner must not be able to predict the day of his execution, he evidently meant that a perfectly logical prisoner would be unable to deduce the date with certainty. A regular, not so logical prisoner has greater freedom. He may have a hunch that it will be a certain day, and may even be right (an unjustified true belief). No choice of execution date is proof against a lucky guess. Assuming the judge's order has any meaning at all, it must prohibit only a rational determination of the date.

For simplicity, use the two-day version of the paradox. Suppose for the sake of argument that the prisoner *can* logically determine that he must be executed on Saturday to fulfill the judge's instructions. The executioner (who is just as bright as the prisoner) can likewise deduce this. He then has no reason to hang the prisoner on Saturday rather than Sunday. See why? The prisoner expects Saturday (that's the premise of this reductio ad absurdum); but even if by some miracle he *isn't* hanged on Saturday, he can deduce it will be on Sunday. That leaves the executioner no reason to prefer one day over another. He is damned if he does hang him on Saturday, and damned if he doesn't.

Consequently, the executioner is equally free to execute the prisoner on either day. That means the prisoner is *wrong* to conclude he will be hanged on Saturday.

If you prefer, you can assume that the prisoner deduces Sunday as the only logical day of execution. This gives the executioner equal reason to hang him on Saturday, and the prisoner is wrong again.

The result is a Gettier situation with a twist. Let's say the prisoner *is* hanged on Saturday. On the surface, the prisoner seems to have been right. The prisoner has a justified true belief. It fails to be genuine advance knowledge, though. The prisoner does not realize the defeater of his belief: that just as good a justification exists for him being hanged on Sunday. As stated above, the assumption that the prisoner *must* be hanged on Saturday leads to the conclusion that he can equally well be hanged on either day. The defeater of the prisoner's belief is the belief itself.

The unexpected hanging is a cautionary tale of deduction. The prisoner deduces that it is impossible to hang him on Sunday and is left only with Saturday. His fatal error is thinking that eliminating the impossible guarantees that something possible will be left over. Sometimes every road leads to contradiction.

The lawyer glimpsed more of the truth in concluding that the order couldn't be carried out. Neither lawyer nor prisoner took the crucial final step: If the prisoner accepts the impossibility of carrying out the order, then the executioner can hang him any day, even the last, and it will be unexpected.

7

THE IMPOSSIBLE

The Expectancy Paradox

YOU ARE THE HEAD of a university psychology department conducting a bizarre experiment on human subjects. Person A sits at a desk, working on a psychological test. Person B sits opposite him, watching his progress. In front of B is a push button. B has been told that pushing the button causes A to receive an excruciatingly painful electric shock (though no permanent injury). Periodically, Professor Jones walks over to A's desk, notes an incorrect answer, and instructs B to press the button.

A is really Jones's confederate. The button is not connected to anything, and A is faking his pain when the button is pressed. Jones is conducting the experiment only to see if B will go along with his instructions to "punish" A. Jones's pet theory is that most people will countenance cruelty if it is approved by an authority figure.

Jones has tried the experiment with ten different B's and eight of them have pressed the button.

Professor Jones is himself unaware of this Kafkaesque irony: *He,* Jones, is the real subject of *your* experiment. You are interested in the "fudge factor"—or "experimenter bias effect"—in psychological experiments. When a researcher *expects* a certain result in a psychological experiment, he is more likely to get that result. Research tends to support the researcher's pet theories—which means something is wrong.

In other lines of research, the experimenter bias effect can be reduced or eliminated. Tests of new drugs are "double-blind" experiments in which some subjects receive the drug, some receive a placebo, and neither subject nor experimenter knows which is which until after the results are in. That prevents the experimenter from communicating enthusiasm for the new drug to those receiving it.

But double-blind controls are almost impossible in some psychological studies. The experimenter necessarily knows what's going on. Take Jones. He expects his subjects to "turn Nazi" and therefore most of them do. Whereas Professor Smith, who believes people are basically decent, did the same experiment and reported that only one person in ten would push the button. The problem isn't conscious fraud; it's subconscious fudging. Both Smith and Jones tend to interpret ambiguous outcomes in favor of their desired conclusion. When Jones tells his subjects to push the button, he is harsher, more imperative than Smith was. Possibly Smith and Jones selected their B's so as to get the wanted outcome. Neither researcher is aware of it, but they are creating self-fulfilling prophecies.

If the experimenter bias effect is widespread, it will have drastic implications for research on human subjects. So you convinced a big foundation to fund your experiment. The subjects of your experiment are other psychologists who have no idea what's really going on. The foundation gave you enough money to fund Jones's experiment, and Smith's, and many others. You don't care one whit about what Jones and Smith and all your other subjects find out in their experiments. The idea is solely to measure any suspicious correlation between a researcher's preconceptions and his results. You have observed many, many psychologists, of varied personalities, running all conceivable types of experiments on unknowing human subjects. The evidence is clear: The experimenter bias effect is both overwhelming and universal. In 90 percent of all cases, the outcome of psychological experiments is whatever the experimenter expected.

And that's the problem. *This* result is exactly what *you* expected. If your study is correct, then the results of psychological experiments on human beings are invalid. *Your* study is a psychological experiment on human subjects. Therefore, your study is invalid. But if your study is invalid, then there is no reason to believe in the experimenter bias effect, and quite possibly your study is valid, in which case it's invalid . . .

Catch-22

"All generalizations are dangerous, even this one," goes an epithet of Alexandre Dumas *fils* with more than a passing resemblance to the above. The "expectancy paradox" is also reminiscent of the paradoxical situation in Joseph Heller's novel *Catch-22:*

> There was only one catch and that was Catch-22, which specified that a concern for one's own safety in the face of dangers that were real and immediate was the process of a rational mind. Orr was crazy and could be grounded. All he had to do was ask; and as soon as he did, he would no longer be crazy and would have to fly more missions. Orr would be crazy to fly more missions and sane if he didn't, but if he was sane he had to fly them. If he flew them he was crazy and didn't have to; but if he didn't want to he was sane and had to.

Compare that with the famous and probably apocryphal story about Protagoras (c. 480–411 B.C.), founder of sophism. Protagoras was the first teacher in ancient Greece to charge money for his lessons. One student of the law struck a bargain with Protagoras: He would pay his tuition upon winning his first law case. If the student lost his first case, he would pay nothing. The student tried to get out of the deal by refusing to accept cases. Protagoras had to sue the student to get his money—and the student defended himself. If the student lost, he would not have to pay, and if he won, he would not have to pay.

(So the story goes, anyway. One may imagine that if the student prevailed on the matter of whether he could postpone taking his first case, Protagoras could immediately demand his fee, and if necessary sue him again for cut-and-dried breach of contract.)

An element common to each of these paradoxes is categories or sets that can contain themselves as members. The crux of the expectancy paradox is that the experiment pertains to the class of experiments on humans, and the experiment is itself in that class. The classic illustration of sets containing themselves as members is Bertrand Russell's "barber paradox": In a certain town, the barber shaves everyone who doesn't shave himself. That is, he shaves *only* those men who don't shave themselves, and *every* such man. Does the barber shave himself? There is no way the barber can live up to his reputation. If the barber doesn't shave himself, he must shave himself, and if he does shave himself, he can't shave himself.

All the above are paradoxes in puzzle's clothing. It sounds at first like there is some resolution to be found, and that once you find it, you'll be able to say, "Aha! This is what would really happen." Then you realize that it's hopeless. No matter what you assume, you end up with an impossibility.

Can Such Things Be?

One common reaction to the type of paradoxes above is to wonder if they are "possible"—that is, if they could ever occur in the real world. In some cases, the answer is certainly yes. Protagoras' lawsuit could have taken place (presenting the judge with a difficult decision); the military could (probably does) have confusing and contradictory rules. A barber could shave *everyone else* in town who doesn't shave himself—leading townsfolk to make Russell's claim about him—though he still could not truly fulfill the claim.

Real experiments have supported the experimenter bias effect (which has even rated an acronym: EBE). In 1963, Robert Rosenthal and K. Fode reported three studies showing a significant effect. Rosenthal and Fode assigned a number of college students to conduct sham experiments on human subjects. The subjects were shown photographs of assorted individuals and asked to decide if the individuals had been "experiencing success" or "experiencing failure." About half the student experimenters were led to believe that their subjects would favor "success" responses; the other group was told to expect "failure" responses. Then the reported results of the sham experiments were compared. Since the sham experiments should have produced the same results each time, the differences were presumed due to the experimenter's expectations. Later studies by Rosenthal further investigated the effect. Rosenthal went so far as to suggest that future experiments on humans might have to be conducted via automated procedures to avoid the taint of bias.

Other researchers were unable to replicate Rosenthal's findings. The matter came to a head in a 1969 issue of the *Journal of Consult-*

ing and Clinical Psychology. The journal published, back to back, a study by Theodore Xenophon Barber and colleagues carefully duplicating Rosenthal's experiments but finding absolutely no evidence for the bias effect; a defensive rebuttal by Rosenthal; and a peevish counterrebuttal by Barber. The undercurrent of irritability sublimated in scientific nitpicking resulted in such deadpan statements as the following (from Barber, in response to Rosenthal's objection that Barber had replicated the experiment at an all-female school): "If Rosenthal is seriously contending the Experimenter Bias Effect is more readily obtained in coeducational state universities than in other types of colleges or universities, he should present data to support the contention."

Subsequent studies have further weakened the case for a widespread bias effect. At least forty studies published from 1968 to 1976 found no statistically significant experimenter expectancy effect, and six others provided but weak evidence of it.

For the expectancy *paradox* to exist in the real world, it would have to be determined that the expectancy effect is both universal and unavoidable. There would be no problem if it's just *some* psychologists who fall victim to the effect. Then the experimenter could be a careful, coolheaded psychologist measuring the foibles of his sloppy colleagues. Just as paradox requires that a Cretan utter "All Cretans are liars," it is necessary that an experiment of a certain type assert the unreliability of *all* experiments of that type.

In reality, it is unlikely that the expectancy effect would be ubiquitous. For that reason, even the actual studies purporting to demonstrate the effect are not necessarily caught up in the maelstrom of paradox.

Okay. But what would it mean if it was indeed determined that the results of all experiments on human beings are invalid, including the experiment that determined that fact? Could that happen?

There is a distinction between falsehood and invalidity. If an experimental result is false, it's false, but if the experiment is merely invalid (through careless procedure, lack of controls, etc.), its results may be true or false. An invalid experiment may support a hypothesis that happens to be true (call this a "Gettier experiment").

In the liar paradox, an assumption of truth leads to falsehood, and an assumption of falsehood leads to truth. But are we talking about the truth/falsehood or validity/invalidity of the expectancy effect experiment here? It is not immediately clear. Let's list all the possibilities, as we might do with a logic puzzle.

(a) Assume the study's results are true. If they are, then psychological experiments on humans can't be trusted. (The study does not purport to show that the results of psychological experiments are invariably *wrong*, just that you can't go by them.) Therefore the experiment can't be trusted either. Its conclusion *could* be true—and in fact *is* true by our assumption—but the study is not valid evidence for it. The study is a Gettier experiment, and this is a possible if ironic state of affairs.

(b) Assume the study's conclusion is false. Then there is no universal expectancy effect. The study's conclusion could be and presumably is false for some other reason. (If the conclusion is false, then the study must also be invalid.) Again, a possible state of affairs.

(c) Assume the study is valid. Then its conclusion is true and the experiment is invalid: a contradiction.

(d) Assume the study is invalid. Then its conclusion may be true or false: no contradiction there.

In short, if someone did a study purporting to show that the experimenter expectancy effect is universal, the tenable criticisms would be that either (a) the conclusion expresses a serendipitous truth that is, however, not justified by the study, which is invalid; or (b) the conclusion is false, and the study invalid; or (d) the study is invalid, period. No matter what, you are forced to conclude that the study is invalid.

But what if a committee of Nobel Prize-winning scientists supervises the study, taking the greatest possible pains to ensure its validity? A system of scrupulous controls, statistical checks, and double checks such as never before seen in an experiment are instituted. Then this undeniably valid study truthfully asserts that all psychological experiments on humans (of which the study is one) are invalid because of subconscious experimenter bias.

This, the heart of the paradox, is the liar paradox with validity substituting for truth. A valid study that asserts its own invalidity just can't be; we have entered the realm of the impossible.

Possible Worlds

One well-known idiom of philosophy is "possible worlds." It is natural to wonder why the world is as it is. Why is there evil? That we even ask that question demonstrates that we can imagine a world without evil, a world much different from the one that exists.

There is reason to believe that the ability to conceive of possible worlds is a fundamental part of human intelligence. All the thousands of choices we make in our lives, momentous and trivial, are acts of imagination. You imagine the world in which you get your car washed this afternoon and the world in which you don't, and decide which you would rather live in.

The first Western writer to use the idea of possible worlds extensively was the German mathematician and philosopher Gottfried Leibniz (1646–1716). Leibniz wondered why, out of all the possible worlds, God chose to create this one. His peculiar answer was that this is in fact the best of all possible worlds. Leibniz imagined that the pain and suffering of the world was at an absolute minimum; any tweaking on the part of the Creator, any attempt to right a wrong here or there, would make things worse globally. This incredible point of view is remembered for inspiring the character of Dr. Pangloss in Voltaire's satire *Candide*. Candide could not see how a world in which the Lisbon earthquake (which killed about 40,000 in 1755) had not occurred would fail to be better than ours.

Possible-worlds philosophy was revived in the 1960s by such philosophers as Saul Kripke, David Lewis, and Jaakko Hintikka. Lest there be any confusion, let's clarify what a "possible world" is. It is *not* another planet out there in space. A possible world is a complete universe unto itself with a past, present, and future. You can talk about the possible world in which Germany won World War II, and even about the year 10,000 A.D. in that possible world. People often use the singular to denote what is actually a class of possible worlds. There must be trillions upon trillions of possible worlds in which Germany won World War II, each differing from one another in some detail. There are, or seem to be, an infinite number. The one possible world that we live in is called the "actual" world.

Even so metaphysical an idea as this has its limits. The concept would not be very useful if any and every effort of imagination constituted a possible world. Most philosophers allow that it is possible to talk about worlds that are not possible worlds.

Though we string these words together, "a world in which 1 plus 1 does not equal 2" does not describe a possible world. Nor could there be a world in which 6 is a prime number; a world in which pentagons have four sides; a world in which the Lisbon earthquake occurs *and* the Lisbon earthquake doesn't occur; or a world in which Abraham Lincoln is taller than Joseph Stalin, Joseph Stalin is taller than Napoleon, and Napoleon is taller than Abraham Lincoln.

(Some dispute this. Though no one has any idea how there could be a world in which 1 and 1 don't equal 2, a dyed-in-the-wool skeptic can doubt our certainty that no such world is possible. Most philosophical discussion of possible worlds takes as a ground rule that our logic at least applies to other possible worlds. If *not,* we are in no position to reason about them.)

How Many Worlds Are Possible?

To say that something is impossible—as opposed to merely false —is to say that there is no possible world in which it could be true. One of the deepest questions in philosophy is how diverse the possible worlds are.

Saul Kripke argued that such facts as "the atomic number of gold is 79" are true in *any* possible world. Most find this hard to accept. It seems easy to imagine a world in which gold's atomic number is 78 or 80 or 17. Quite possibly, you've lived your whole life not knowing or much caring what gold's atomic number is. Envisioning a different atomic number for gold appears little different from imagining a world in which your phone number or license plate is different. But is it?

The properties of elements can be predicted from their position in the periodic table. Gold falls beneath silver and copper in the table and resembles them in many ways. It is a dense, soft, unreactive metal that conducts electricity very well. Were gold's atomic number even one more or less, it would occupy a different position and would be expected to possess different properties.

Suppose gold's atomic number was 78. It would fall beneath nickel and palladium in the table and resemble them. It would still be a dense metal, but its properties ought to be more like platinum (which in fact has atomic number 78). Would "gold" that resembles platinum in all respects be gold at all?

You could contend that the other elements would be shifted one atomic number down in the periodic table so that gold could still occupy the same relative position. Gold would be element 78, platinum would be element 77, and so on. But then you'd drop an element at the beginning of the periodic table. The dropped element would be hydrogen, which makes up stars and is by far the most common element in our universe. A universe without hydrogen would be so different that we are unable even to guess how different it would be.

To a chemist, Kripke maintained, the elements have properties that follow more or less inexorably from their atomic numbers. The idea of a world in which helium is not an inert gas is not so much different from the idea of a world in which 2 is not $1 + 1$. Deciding if a world is possible is trickier than it looks!

There may come a day when our knowledge of physics is as complete as the current state of chemistry. It is conceivable that the properties of electrons, quarks, and photons have the same underlying justification as the properties of chemical elements. "Superstring" theories try to provide just that. If they are right, many exotic worlds that seem to be possible (a world in which protons are more massive than neutrons; a world in which electrons are the size of golf balls) may actually be ruled out. Physicists have even speculated that the actual world is the only one possible. The laws of physics and even the initial state of the world may be preordained with a logical rigor we can scarcely imagine.

Paradox and Possible Worlds

When we say that "This sentence is false" is paradoxical, we mean that there is no possible world in which that sentence accurately describes itself. The situation can be broken into two parts: (1) if the sentence is true, then it is false; and (2) if the sentence is false, then it is true. We are free to imagine worlds in which the sentence is true or false, but both alternatives lead to contradiction.

Jaakko Hintikka defined knowledge via possible worlds. To increase one's knowledge is to decrease the number of possible worlds compatible with what one knows. For instance, everything we know is compatible with there being life in the Alpha Centauri star system, and everything we know is compatible with there *not* being life in Alpha Centauri. Such is our ignorance that we cannot distinguish the real world from a merely possible world identical to ours in every way except in whether there is life in Alpha Centauri. If and when we find out whether there is life in Alpha Centauri, one set of possible worlds will be ruled out.

Scientific discovery decreases the number of compatible possible worlds. It is natural to ask how far this process may be carried. In Hintikka's view, total knowledge would mean paring away all the possible worlds until just *one* remains—the actual world.

Notice the slender distinction between omniscience and paradox. To someone utterly ignorant, the number of possible worlds compatible with his knowledge is infinite. To someone gifted with total knowledge, the number of possible worlds is narrowed to one. What if the field was narrowed to *zero?* That would be the predicament of someone who has discovered that no possible world is compatible with what he knows. His set of known facts includes a contradiction. The best paradoxes seem to prove that *this* is not a possible world.

In the essay "Avatars of the Tortoise," Borges speculated that paradoxes were clues to the unreality of the world:

> Let us admit what all idealists admit: the hallucinatory nature of the world. Let us do what no idealist has done: seek unrealities which confirm that nature. We shall find them, I believe, in the antinomies of Kant and in the dialectic of Zeno.
>
> "The greatest magician (Novalis had memorably written) would be the one who would cast over himself a spell so complete that he would take his own phantasmagorias as autonomous appearances. Would this not be our own case?" I conjecture that this is so. We (the undivided divinity operating within us) have dreamt the world. We have dreamt it as firm, mysterious, visible, ubiquitous in space and durable in time; but in its architecture we have allowed tenuous and eternal crevices of unreason which tell us it is false.

The Paradox of the Preface

We have all seen those overly self-effacing prefaces in which the author (after thanking spouse and typist) takes responsibility for the "inevitable" errors. You've probably wondered why, if the author is so sure there are errors, he doesn't go back and correct rather than acknowledge them. Inspired by these disclaimers, D. C. Makinson developed the "paradox of the preface" (1965). Related to both the expectancy paradox and the unexpected hanging, the paradox of the preface "proves" that there is no such thing as nonfiction.

An author writes a long book he believes to be nonfiction. It makes many statements that he has carefully checked. A friend reads the book, shrugs, and says, "Any book *that* long contains at least one error." "Where?" the author demands. The friend avers that he didn't catch any error, but still, virtually all long nonfiction books do have an error or two. Reluctantly, the author agrees. "Then," says the friend, "your readers are not justified in believing *any* statement in your book."

"Look," the friend says. "Pick a statement." He opens the book at random and points to a declarative sentence. "Ignore this statement for the moment. I'll put my finger over it so you can't see it. Do you believe that every *other* statement in the book is true?"

"Of course. I wouldn't have made the statements unless I believed them, and had good justification for believing them."

"Right. And you agree that the book must contain at least one error, even though neither you nor I have spotted it. If you believe the book contains at least one error, and further believe that every statement other than this one is true, then you *must* believe that this statement I am covering with my finger is false. Otherwise your beliefs are self-contradictory. And I just chose this statement as an example. I could have chosen any statement and said the same thing of it. You can't legitimately believe that any of the statements in your book are true," the friend concluded.

Not wanting to mislead his readers, the author wrote a preface to the book warning: "At least one of the statements in this book is false."

If the book contains one or more errors, then the prefatory statement is accurate. If the book exclusive of the preface contains no errors, then the prefatory statement is in error. Then there *is* an error in the book after all, and the prefatory statement is correct. But if the prefatory statement is correct, then there is no error and it is wrong . . . A series of errata sheets inserted in later editions of the book did little to resolve the matter!

Must Justified Beliefs Be Compatible?

Many real prefaces do admit errors. Kurt Vonnegut, Jr.'s novel *Cat's Cradle* is prefaced with the statement: "Nothing in this book is true." This is not Makinson's paradox of the preface but a more directly contradictory relation. Insofar as Vonnegut's book is a work of fiction, the prefatory statement is accurate *except* as concerns the prefatory remark itself. The preface, presumably coming from the real Kurt Vonnegut and not a fictional character, is nonfiction. Speaking of itself, it creates a liar's paradox.

The paradox of the preface also brings to mind mathematician William Shanks's tragic lifework of computing pi and making an error in the 528th decimal place that invalidated all subsequent work. Imagine you are writing a book called *The Digits of Pi.* On page one you write: "The first significant digit of pi is 3." Each succeeding page asserts the next digit in pi's decimal expansion. You derive the digits by hand calculation. You are a competent mathematician using an accepted algorithm. Therefore you have justification for believing each and every digit you derive.

By the time you get to the 1000th digit, you realize that, very likely, you have made at least one error in your math. Oops! Now things are much worse than in Makinson's paradox. Calculating each new digit depends on the values of the previous digits (as in long division). You do not determine the 1000th digit of pi directly; you must first determine the 999th digit, and before that the 998th digit, and so forth. Any error in determining a given digit will render all succeeding digits invalid. It is like setting up a line of 1000 dominoes: When the 307th domino falls to the right, all the dominoes after it follow. If you've made at least one error in the first 1000 digits, the 1000th digit must be wrong.[1] So, very likely, is the 999th digit, the 998th digit, and a long string of the digits before them.

Like the expectancy paradox, the paradox of the preface questions the role of deductive reasoning in situations involving inductive probabilities rather than certainties. Since probabilities rather than certainties are the lot of the scientist, it deserves a thoughtful response.

Our worldview is a set of beliefs, mostly justified and mostly true (so we think, anyway). The paradox of the preface asks whether it is possible to have justified beliefs that are logically contradictory. Note the paradox within the paradox. The author has a set of beliefs (that each statement, considered individually, is true, plus the belief that the book contains an error) that contains a contradiction. Suppose the book proper makes 1000 distinct assertions, which are mutually compatible. The claim of the preface ("At least one statement in this book is wrong") is the 1001st assertion. This yields the most recherché of contradictions, for any 1000 of the 1001 statements are logically consistent, even though the complete set of 1001 is self-contradictory.

Probability enters more explicitly into the related "lottery paradox" of Henry E. Kyburg, Jr. (1961). No one who buys a lottery ticket can reasonably expect to win; the odds against it are too high. Yet everyone's expectation of not winning conflicts with the fact that somebody *will* win. In practice the chain of suspect reasoning goes a step further. Many lottery players justify their wagers with "Someone has to win, so why not me?"—a fallacious rationale that is echoed in state lottery advertising. Kyburg felt that his paradox shows that one's set of justified beliefs *can* be logically inconsistent.

One subtext of Makinson's and Kyburg's paradoxes is the way a large number of beliefs may conceal contradictions. A single state-

---

[1] There is a 1 in 10 chance it is correct, and a Gettier counterexample!

ment can subtly introduce contradiction in a set of millions. Take this sorites:

1. Alice is a logician.
2. All logicians eat pork chops.
3. All pork-chop eaters are Cretans.
4. All Cretans are liars.
5. All liars are cabdrivers.

    .
    .
    .

999,997. All Texans are rich.
999,998. All rich people are unhappy.
999,999. All unhappy people smoke cigarettes.
1,000,000. Alice doesn't smoke.

The dots signify that premises 6 through 999,996 are additional statements of the type "all X's are Y's," so that one may ultimately conclude that all logicians are cigarette smokers, and from that, that Alice smokes. That contradicts the 1,000,000th premise, so the set is unsatisfiable (self-contradictory).

Nothing so remarkable there. The surprising thing is that removing any *one* premise makes the set satisfiable. Strike out premise 4. Then you can conclude that Alice is a Cretan, that all liars smoke, and that Alice does not smoke (and thus isn't a liar).

In this example, the premises are in a neat order to facilitate seeing the contradiction. If the million premises were shuffled into random order, it would be a more arduous task to see that the set was self-contradictory. If some of the statements were more complicated, it would be more difficult yet. A set of beliefs is like the Borromean rings, or a mechanical puzzle where the removal of one piece causes all the others to fall apart. The influence of each assertion can "ripple out" and affect the whole set.

Pollock's Gas Chamber

When embroiled in paradox, one's impulse is to give up one or more of the original assumptions that have led to contradiction. The question is, how do you decide which belief to give up? John L. Pollock resolved the paradox of the preface through rules of confirmation he illustrated with this thought experiment:

A room is occasionally filled with a poisonous green gas. To warn those who might want to enter, the room is supplied with a warning system. The system (which was designed by a committee) works like this: A warning light is visible through a window in the door to the room. The light is green (for "go") when it is safe to enter. It is white (the color of death in some Asian countries) when the room contains the deadly gas.

Unfortunately, the system is worthless because the green gas makes the light look green when it is actually white. The light *always* looks green, gas or no gas. The committee has remedied this horrible deficiency by mounting a closed-circuit television camera just inches from the warning light. The video signal goes to a color monitor outside the room. The monitor accurately reproduces the color of the warning light, whether the room contains the gas or not. A sign on the door warns the public to ignore the apparent color of the light through the window and instead consult the television monitor.

Pollock's kludgy warning system is an allegory of our imperfect knowledge of the world. The light is green or white; we do not know which. It looks green through the window. That is prima facie evidence for believing it to be green. The light looks white on the TV screen. That is reason for believing it to be white. But if it is green it cannot be white, and vice versa. We must give up one of these initially credible suppositions.

Pollock notes that there is more than one way of rejecting a belief. You might say, "The light looks green through the window. I know from experience that most windows are made out of colorless glass that does not distort colors, and that air is colorless too. Therefore, the light's appearance through the glass is justification for believing that it *is* green. If it's green, it can't be white. So it's not white."

Of course, you could just as easily say something like this: "The light is white on the television monitor. Things usually are the color they appear to be on color TV—that's the whole point of having color TV. Therefore, the light's appearance on the monitor is a good reason for believing that it's white. If it's white, it can't be green. So it's not green."

We have a mini-paradox, in that reasoning from a small set of observations leads to contradiction. Each line of reasoning rebuts the other in what seems to be the strongest way possible.

The resolution is obvious. The light must really be white, as it appears on the TV screen. But we *aren't* using the second argument above. The second argument is no stronger than the first—maybe slightly weaker. (When what you see on TV conflicts with what you

see directly, you probably prefer the evidence of your own eyes.)
There is another argument for the light being white, symbolized by
the sign on the door.

All empirical beliefs are defeasible. It is always possible that you
could learn something (a defeater) to invalidate a belief. There are
two types of defeaters, *rebutting* defeaters and *undercutting*
defeaters.

A rebutting defeater flatly asserts that the belief is wrong. Learning of a colony of white ravens in the Copenhagen zoo would be a
rebutting defeater of the hypothesis that all ravens are black. You
would still have all the evidence you ever had for this hypothesis
(all the sightings of black ravens) and it would still "count," yet you
would be forced to admit that the hypothesis is false.

An undercutting defeater demonstrates that the evidence for the
belief is invalid. Learning that you are actually a brain in a vat
would be an undercutting defeater for *everything* you believe about
the external world. An undercutting defeater puts the "evidence"
for a belief in a new light, and shows that it cannot be used to justify
the belief. The belief might still happen to be true, but the supposed
evidence is bad.

It sounds like the rebutting defeater is the stronger of the two.
Actually, said Pollock, undercutting defeaters take precedence over
rebutting defeaters. It is like the difference between an interesting
debate and a boring one: In the latter, the opponents alternate telling each other they're wrong; in the former, they say *why* their
opponent is wrong.

The empirically justified conclusions about the light (that it is
green based on its appearance through the window; that it is white
based on its appearance on the TV screen) are rebutting defeaters of
each other. The situation is resolved only through the sign, an undercutting defeater. By explaining that the light may look decep-
tively green when seen through the green gas, it gives us reason to
throw out one belief and keep the other.

This principle of the dominance of undercutting defeaters helps
make sense of most of the paradoxes of this chapter (and the unexpected hanging as well). The argument of the author's friend in the
paradox of the preface is a rebutting defeater of the singled-out
statement. It says the statement is *wrong,* without saying *why.* The
reasoning is quite external to the statement. In fact, the content of
the covered (and unread) statement never enters into it.

The author could cite an undercutting defeater for the friend's
argument. The friend's reasoning rests on the belief that the book

contains an error. Although there may be excellent empirical evidence for that belief (finding mistakes and typos in other books), it would surely be undermined were it known for a fact that all the book's statements other than the one the friend singled out are correct. Then the only way the book could contain an error would be for the covered statement to be wrong—and there is no reason to believe that it is any more likely to be wrong than the other statements. When push comes to shove, you should go with the undercutting defeater.

The paradox of the preface is a facetious paradox. We knew all along that the friend's argument was wrong; the riddle was to say exactly why. The expectancy paradox is a tougher nut to crack. Applying Pollock's principle to it yields the following resolution (not necessarily the last word):

An argument that an experimental result is false is a rebutting defeater; showing that an experiment is invalid is an undercutting defeater. In case of conflict, Pollock would have us favor a demonstration that the experiment on the expectancy effect is invalid rather than that it is false.

Take the strong version of the paradox, where the blue-ribbon committee of famous scientists has supervised the experiment, and we therefore are assured of the experiment's validity. The rebutting defeater is this: If the results are true, then the experiment must be invalid. But since we *know* the experiment is valid (thanks to the expert supervision), the results must not be true (by *modus tollens*).

The undercutting defeater goes: If the experiment is valid and true, then our subconscious expectations have compromised the experiment. Regrettably, we conclude that the experiment is invalid. (For what it's worth, this seems the more reasonable of the two positions.)

And finally, for the unexpected hanging (which resembles the paradox of the preface in the plurality of days/statements): The prisoner's reasoning rebuts the possibility of his being hanged on each of the seven days of the week. This set of beliefs creates its own undercutting defeater, for the executioner, aware of the prisoner's beliefs, can hang him any day. Favoring the undercutting defeater gives us Quine's position that the prisoner is wrong.

You might wonder when you can conclude that something is established beyond all doubt. The answer is: never. This is the trouble with accepting indefeasibility as a fourth criterion of knowledge. No belief is immune from defeaters—not even a belief that *is* a defeater.

A watchman approaches and inspects the monitor outside Pollock's gas chamber. "Kids!" he mutters. "They think it's some big joke to play with the controls of this thing! Just wait till someone gets *killed*, then they'll do something," he grouses, twiddling the dial and turning the image of the light bulb a vivid green.

8

INFINITY

The Thomson Lamp

THE "THOMSON LAMP" (after James F. Thomson) looks like any other lamp with a toggling on-off switch. Push the switch once and the lamp is on. Push it again to turn it off. Push it still another time to turn it on again. A supernatural being likes to play with the lamp as follows: It turns the lamp on for $1/2$ minute, then switches it off for $1/4$ minute, then switches it on again for $1/8$ minute, off for $1/16$ minute, and so on. This familiar infinite series ($1/2 + 1/4 + 1/8 + \ldots$) adds up to unity. So at the end of one minute, the being has pushed the switch an *infinite* number of times. Is the lamp on or off at the end of the minute?

Now, sure, everyone knows that the lamp is *physically* impossible. Mundane physics shouldn't hamper our imaginations, though. The description of the lamp's operation is as logically precise as it can be. It seems indisputable that we have all the necessary information to say if the lamp would be on or off. It seems equally indisputable that the lamp has to be either on or off.

But to answer the riddle of the Thomson lamp would be preposterous. It would be tantamount to saying whether the biggest whole number is even or odd!

The Pi Machine

The unease is greater yet with the "pi machine." This amazing device looks like an old-fashioned cash register. Switch it on, and the pi machine swiftly calculates the digits of pi (the length of a circle's circumference when its diameter is 1). As has been known since classical times, pi is an *endless* string of digits: 3.14159265 . . . The pi machine telescopes this infinity by computing each successive digit in only half the time of the preceding one. As it determines each digit, the numeral pops up in a window at the top of the machine. Only the single most recently calculated digit is visible at any instant.

If the first digit requires 30 seconds of computation, the machine will calculate all the digits of pi in a minute.[1] What's more, at the end of the minute, the machine will be displaying the actual, bona fide "last" digit of pi! Of course, the latter is pure moonshine, for pi has no last digit.

Rounding out this trio of impossible machines is the "Peano machine." This is something like an automated slide whistle. The whistle is calibrated like a ruler. One end is labeled "0," the other end "1." A plunger travels from the "1" end to the "0" end at constant velocity in one minute's time. As the plunger passes a point whose inverse is a whole number, a pair of mechanical lips announces the number. The machine's voice gets higher as the plunger moves down, allowing it to recite ever faster and faster.

For instance, at the very start of the minute, the plunger is at 1, and 1/1 is 1. The machine recites "One" in a rich baritone. Thirty seconds later, the plunger is at 0.5. The inverse of that is 2, and the machine says "Two" (now in tenor). Ten seconds later comes a contralto "Three." Five seconds after that, a soprano "Four."

Toward the end of the minute, the recitations come fast and furious. The crescendo becomes too high-pitched for anyone to hear.

[1] Running the pi machine simultaneously with the Thomson lamp in an otherwise dark room affords a stroboscopic view of the odd-numbered digits of pi.

Dogs whine and paw the ground frantically a few moments . . . then even they can't hear the machine. By the end of the minute, the name of every natural number will have been spoken.

Zeno's Paradoxes

The infinite, a symbol of the vast world which cannot be fully grasped, is a common motif of paradoxes. Frequently, the paradox has the infinite impinging upon and threatening the complacent everyday world.

Among the oldest paradoxes of the infinite are those attributed to Zeno of Elea (lived fifth century B.C.). Zeno described his paradoxes in a book (written about 460 B.C.?) that has been lost. The paradoxes are known to us only via the often abbreviated accounts of them by other ancient writers. Zeno was a curmudgeon who delighted in demonstrating that time, motion, and other commonplaces cannot exist. His best-known paradox goes like this: Swift Achilles races a tortoise; the tortoise has a head start of, say, a meter. In order to overtake the tortoise, Achilles must run the meter to the tortoise's starting point. In the time it takes him to do that, the tortoise will advance a shorter distance, 10 centimeters. Now Achilles must run 10 centimeters to gain the lead. Meanwhile the tortoise pulls out 1 centimeter ahead. This analysis can go on forever; the tortoise's lead dwindles but never does Achilles overtake it.

Zeno denied the reality of infinite series or quantities. He felt that if you could show that something involved an infinity, you proved it couldn't be. To modern minds, some of Zeno's arguments are less compelling. Zeno is apt to come off as a mathematical crank who never got the hang of infinite series. The series of intervals Achilles must run adds up to a finite total: 111.1111 . . . (or 111 and 1/9) centimeters. The "infinity" is more in Zeno's analysis than in the physical situation.

A more puzzling invention of Zeno's is the arrow paradox. An arrow flies through the air. At any point in time the arrow is motionless. The instantaneous arrow is like a still photo of the arrow or one frame of a movie of it. Time is made up of an infinity of these instants, and in each instant the arrow is stock-still. Where is the arrow's motion?

The arrow paradox repays further thought. Put it in a more modern context. We have an arrow of atoms; it moves in the space-time of relativity and is measured in an inertial frame of reference. Even in that context, "an instant of time" has something of the informal meaning it did to Zeno. We still believe in cause and effect (except at the quantum level: negligible here?), that the future is determined by the present, and the present by the past. Now, in that frozen instant of time, what distinguishes a moving arrow from a stationary one? It would seem there must be some information attached to a moving arrow that identifies it. Otherwise, how does it "know" to jerk forward in the next instant?

More within the scope of this book are the contemporary "infinity machine" paradoxes above. Inspired by Zeno, they question knowledge rather than kinematics. Nor does the modern concept of an infinite series do anything to dispel them. The operation of each machine is a supertask, an infinity of action that, while perhaps impossible, can be described unambiguously. In each case, the supertask promises us a glimpse of the Medusa—something that seems unknowable.

The practical-minded may question the point of infinity machines. Philosophic discussion of supertasks is like a doctor looking for a cure for a nonexistent disease. However, there is an analogy between supertasks and certain real-world processes. The peculiar status of questions that can be answered only through an infinite (or "practically infinite") series of discrete actions is worth exploring.

Building a Thomson Lamp

Some discussion of infinity machines has focused on their nuts-and-bolts operation. Although their practicality seems irrelevant, a slightly more detailed analysis may point up logical difficulties. Adolf Grünbaum analyzed all three machines.

One objection to the Thomson lamp might be that a light bulb can't be switched on and off infinitely quickly. Past a certain point in the process, the filament won't have time to heat up fully when the current is on, or cool off when the current is off. Possibly the filament will remain half incandescent throughout the last moments.

Besides, everyone knows that switching lights on and off is a good way to burn out bulbs. The Thomson lamp's bulb would burn out for sure.

Grünbaum argued that these issues are not crucial. The riddle is, will the light be on or off at the end of the minute? After the minute is up, you can always unscrew the burnt-out bulb and pop in a fresh one. Will it light up?

The real problem is with the switch. The on-off button in a Thomson lamp evidently travels a distance x each time it is turned on or off. Therefore the button must travel an infinite distance in a finite time. To mention just one physical objection: Near the end of the minute, the button must be traveling faster than the speed of light, which is impossible.

It isn't essential that the button traverse an infinite in-and-out route—after all, it's not going anywhere. Grünbaum and Allen Janis tinkered around a bit and came up with a modified Thomson lamp that would be more plausible.

Picture the button as a vertical cylinder with an electrically conducting base. When the button is fully depressed, its base makes contact with two exposed ends of the circuit. The current flows through the base and lights the bulb.

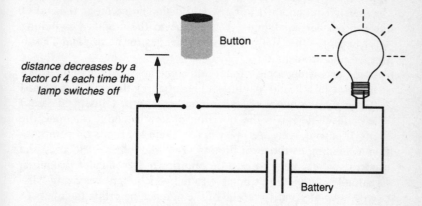

Button

distance decreases by a factor of 4 each time the lamp switches off

Battery

Whenever the lamp is supposed to be on, the button is resting on the open circuit. Whenever the lamp is supposed to be off, the button makes an up-and-down excursion at unvarying speed. Each time, the button ascends only as far as is permitted by the time available and the fixed speed.

During the first 30 seconds the button is flush against the circuit ends and the bulb is lighted. During the next 15 seconds, the bulb is off. The button travels up for 7.5 seconds, then down for 7.5 seconds. Then it stays put for another 7.5 seconds, closing the circuit and lighting the bulb again. Next, the button ascends for 1.875 seconds and descends for 1.875 seconds, keeping the bulb dark for 3.75 seconds.

The button ascends and descends an infinite number of times, but each time it ascends only a quarter of the previous distance. It is like a not very bouncy ball. The total distance traversed is as tidily finite as the total time for the operation. The speed is constant and much less than the speed of light.

Unfortunately, Grünbaum and Janis's modified Thomson lamp isn't completely housebroken. Its back-and-forth motion must also entail arbitrarily great acceleration and deceleration. Presumably, infinite acceleration is easier to swallow than infinite velocity. Still . . . any physical object can stand only so much acceleration. At some point the acceleration would destroy the button as surely as if you smashed it with a hammer.

A worse problem with the modified lamp is that there is no question whether it would be lighted at the end of the minute. The base of the button remains ever closer to the open circuit and finally ends up right on top of it (just as a bouncing ball ends up flush against the floor). The modified lamp would definitely be on at the end of the process. Unsatisfyingly, this is due to the modified switching arrangement. What if any relevance this has to the original Thomson lamp is problematic.

Other problems, some similar and some not, face the builder of a pi machine or a Peano machine. (The latter, by the way, was named by Grünbaum in honor of Italian number theorist Giuseppe Peano.) The pi machine raises the question of how it could calculate the digits of pi so quickly. As we will see, there are limits on computation as binding as those on motion. To avoid infinite velocities, the numerals would have to pop up and down ever smaller distances. Eventually you wouldn't be able to tell which digit was being "displayed." An alternative model of the pi machine prints the digits of pi in a surreal typeface where every digit is half as wide as its predecessor. The complete printout fits on an index card, but not even the most powerful electron microscope will reveal the last digit.

A problem unique to the Peano machine is the ever-lengthening names of numbers. It takes a long time to rattle off the name of a hundred-digit number. Janis suggested that the machine dispense with English names and "whistle" the numbers in a code that pairs every number with a tone of a certain frequency.

The energy required to produce a sound depends on the frequency (pitch) and amplitude (loudness). The amplitude of the tones must decrease in step with the increase of the frequencies to avoid an infinite requirement of energy. By the end of the minute, the volume control on the mechanical lips will be down to zero. You couldn't hear the last whistle, even if you were capable of hearing tones of infinitely high pitch.

Notice this: The attempt to render any of the three infinity machines more physically realizable leads to the conclusion that the result would be invisible (or inaudible). Many philosophers think there is something fishy about infinity machines, supertasks, and "facts" knowable only through supertasks.

Geometric Progressions

The literal infinite is inconceivable, but that which verges on infinity is everywhere. An Indian legend tells how King Shirim was bested by his grand vizier, Sissa Ben Dahir, the inventor of chess. The king was so grateful for the new game that he offered to reward Sissa with a gold piece for each of the 64 squares of a chessboard. The vizier politely declined and asked for an alternative reward. He asked the king to place a single grain of wheat on the first square of the chessboard, then place two grains on the second square, four grains on the next square, and so on, doubling the number of grains on each succeeding square until each square of the chessboard was covered.

The king was taken aback by the modesty of Sissa's demand, and called for a bag of wheat. The grains were counted out carefully as Sissa asked. When the king's servants came to the 12th square, they had trouble fitting all the grains on the square, so they continued by placing the vizier's allotted grains in a heap to the side of the chessboard. The king saw to his amazement that the bag of wheat ran out before the 20th square could be accounted for. He sent for more bags of wheat . . . and finally gave up. Not all the wheat in his kingdom, or in India, or the *world* would fulfill Sissa's request.

The moral, oddly mathematical for a folktale, is never to underestimate a geometric progression. The king's original offer of gold pieces was directly proportional to the number of squares on the board. Had Sissa designed the chessboard to have 81 squares or 49 or some other number, it wouldn't have made all that much difference to the king's grand gesture. What are a few gold pieces more or less to the wealth of a king?

Geometric progressions, however, grow beyond all worldly limits of wealth or anything else. The fact that the unit of the vizier's request, a mere grain of wheat, was so trifling compared to a gold piece scarcely changed matters.

Let's see how many grains would be required to satisfy Sissa's request. It is $1 + 2 + 4 + 8 + \ldots$ Another way of writing this is $2^0 + 2^1 + 2^2 + 2^3 + \ldots 2^{62} + 2^{63}$. (The series ends with 2^{63}, not 2^{64}, because the first square has 2^0 or 1 grain.)

The sum of a series of consecutive powers of 2 is always 1 less than the next greater power of 2. That is, $2^0 + 2^1 + 2^2 (= 1 + 2 + 4)$ is one less than $2^3 (= 8)$. The total number of grains of wheat needed works out to $2^{64} - 1$. That equals 18,446,744,073,709,551,615.

There are something like 100 million grains in a ton of wheat, so this amounts to about 200 billion tons. The current annual production of wheat is only about 460 million tons. The king owed Sissa about four centuries' worth of present world wheat production. Obviously wheat production was a lot less back then. (Just how far back is uncertain, for the date of the invention of chess is unknown. Like baseball, it went through several incarnations, and whether a historic Sissa Ben Dahir existed is also unknown.)

The Malthusian Catastrophe

Thomas Malthus's famous tract was motivated by the realization that the world's population was increasing geometrically but that food production was increasing at only an arithmetic rate. Malthus had reason to believe that the number of new acres open to agriculture each year was approximately fixed. Thus the food supply could grow something like this: 100, 102, 104, 106 . . . On the other hand, the rate of population growth (depending mainly on the number of babies born each year) grows with the size of the population itself. The more people of child-bearing age, the more babies. The world's population tends to double every so many years, growing like this: 1, 2, 4, 8, 16, 32 . . . This, like Sissa's reward, is a geometric progression. It is destined to outstrip the available food supplies, resulting in global famine, Malthus warned.

"Geometric" is a poor term for these series, the analogy to geometry being weak and confusing. A better term is "exponential," from the word "exponent." Exponential growth is characteristic of living organisms. Whether it is a bacteria culture or the population of the human race, the number of new individuals is proportional to the total number. Savings accounts with compound interest grow exponentially—a circumstance that evidently has something to do with the fact that it is living organisms that lend and borrow, create an economy that grows exponentially, and trade in currency that inflates exponentially.

Exponential growth may be described by a simple mathematical function. A function is a procedure that transforms one number into another. Think of it as a special key on a pocket calculator. You punch in a number, press the key, and get a new number. The square root function (which is a key on many calculators) gives a number that when multiplied by itself, yields the number punched in. If you enter 36 and press the square root key, you get 6.

A function does not have to be one of the functions you find on calculators. Any clear and exact procedure for constructing new numbers from old will do. You can define a function as 67 times n plus 381 (for any number n), and it will be a valid function. A function is often described in an equation like this:

$$f(n) = 67n + 381$$

"$f(n)$" is read "the function of n."

Just as it is natural to wonder which animal is the largest or the fastest, mathematicians have wondered which functions are the largest or fastest-growing. Some functions overtake other functions. A function is said to be "bigger" or "faster-growing" than another if its values are always greater *for big enough values* of n. Notice that there is no limit on "big enough." If function A is $A(n) =$ 1,000,000,000,000,000 and B is $B(n) = n$, B will take a long time to catch up to A. For any n greater than 1,000,000,000,000,000, though, $B(n)$ will be greater than $A(n)$. B is therefore faster-growing than A.

Neither of these functions is remarkable. Any constant function —where $f(n)$ equals a fixed value—will eventually be surpassed by any function that is proportional to n. It is also evident that any function proportional to n^2 will outgrow either type of function. A function proportional to n^3 will ultimately grow bigger yet, and likewise for functions in n^4, n^5, n^6, and so on.

A *polynomial* is an expression, such as $n^3 + 8n^2 - 17n + 3$, combining powers of a variable. A polynomial describes a function, and the relative rate of growth of a polynomial function is, loosely speaking, a matter of the highest power. The function $n^3 + 8n^2 - 17n + 3$ grows much larger than any function whose highest power is n^2; in turn it is surpassed by a function containing n^4 or a higher power.

Many functions grow yet faster. Malthus's pessimism was founded on the fact that exponential functions grow faster than any polynomial functions. In an exponential function, you set a certain constant number to the power n (rather than n to a constant power). $f(n) = 3^n$ is an exponential function. This means 3 multiplied by itself n times. If n is 2, 3^n is 3^2 or 9. When n is 1, the result is just the base (3 in this case), and when n is 0, it is defined to be 1 no matter what the base. So the values of 3^n for 0, 1, 2, 3, 4 . . . are 1, 3, 9, 27, 81 . . . Each successive value is 3 times bigger than its predecessor. The higher the base, the faster the function grows. The successive values of 10^n are 10 times bigger, and the values of 1000^n are a thousand times bigger.

In complexity theory, the difficulty of problems is most commonly measured by the time required to solve them. It goes without saying that not all persons work at the same rate. Neither do computers. Just as important, there can be more than one algorithm that solves a problem, and some algorithms are faster than others. The differences in time requirement for various classes of problems are so vast, however, that they dwarf the differences in calculating speed between computers (or people).

In particular, some problems can be solved in "polynomial time" while others require "exponential time." This means that the time required to solve a problem can be expressed as a polynomial (or exponential) function of the size or complexity of the problem. Usually, a problem requiring polynomial time is practical to solve. A problem requiring exponential time is often hopeless. Infinity machines may be chimeras, but exponential-time problems are real and ubiquitous. Solving them can require a "practically" infinite number of steps, even in a finite universe.

The distinction between polynomial-time and exponential-time problems, and the way it relates to paradox, will be explored in the next chapter. For now let's look at two paradoxes that question the infinity of space and time.

Olbers's Paradox

In 1826, German astronomer Heinrich Wilhelm Olbers realized that something is wrong with the universe. Among the sciences, astronomy cannot ignore infinity. Either the physical universe is endless or it is finite. Neither possibility is easy for most people to accept.

"When I consider the small span of my life absorbed in the eternity of all time, or the small part of space which I can touch or see engulfed by the infinite immensity of spaces that I know not and that know me not, I am frightened and astonished to see myself here instead of there," Blaise Pascal wrote. A finite universe may be even harder to believe in. The mind rebels at imagining how space can end.

The unease was not new. Greek philosopher Lucretius felt he could prove the infinity of space with this argument: If space is finite, it has a boundary. Let someone go to that ultima Thule and throw a dart past it. Either the dart will fly past the boundary or something will stop it—something that must itself be just beyond the boundary. Either way, there is something beyond the boundary. This demonstration can be repeated any number of times to push back the putative boundary ad infinitum.

Most astronomers of Olbers's time took the infinity of space for granted. Olbers's reaction was a haunting fantasy that bears his name. Assume that the universe is infinite and that the stars (and galaxies, which Olbers and his contemporaries didn't know about) extend out in all directions forever. In that case, a straight line extended in any direction from the earth must hit a star.

The line may, of course, have to be extended billions of light-years. The point is that in an infinite universe of scattered stars, the line must hit a star *eventually*. This is no more exceptionable than the observation that if you spin a roulette wheel long enough, any given number must come up.

The sun is a star, the only one in our sky of perceptible breadth. Were the sun ten times farther away, it would cover a mere one-hundredth of the apparent area it does and be a hundredth as bright (this by the long-established formula for the attenuation of light). Were the sun a million times farther away, it would be a trillion times dimmer and its disk in the sky would be a trillion times smaller. Notice that the brightness *per area of sky* would be the same. It would be the same no matter what distance the sun was from the earth. On this simple fact, Olbers realized, rests a paradox.

The other stars are pinpricks in the bowl of night, but those pinpricks are (on the average) as dazzling as the sun's surface. Light travels in a straight line. If a straight line extending from the earth hits a star, we see that star's light. And if *every* straight line extended from the earth hits a star, the entire sky should consist of the overlapping disks of stars—each as blindingly bright as the solar disk—fusing into an all-enclosing celestial sphere. It should be as if the sun were a hollow sphere, with us in the middle. There should be no such thing as a shadow, including the shadow we call night.

From this panoramic sun there would be no respite. You might think that dark objects would block some of the starlight from our gaze. But nothing could be dark under these circumstances. All objects must absorb, transmit, or reflect light (usually a combination of all three). Anything that absorbs light (the moon, cosmic dust, this book, your eyelids) should heat up until it is the same average temperature as the stars themselves. Then it should shine with the same intensity. Anything that transmits light perfectly (an ideal pane of glass) is by definition transparent and provides no shade. Objects that reflect light (a mirror) should cast back a glare identical to the background and be invisible.

This reasoning, which is clearly wrong, is Olbers's paradox. Olbers got the credit, but he was not the first to think along these lines. The idea had been kicked around for centuries, attracting the attention of Thomas Digges, Edmund Halley, and Edgar Allan Poe, among others. Like the infinity machines, the paradox apparently supplies a tantalizing cosmic truth (whether the universe is infinite) in short order.

Against Plurality

Looking through the other end of the telescope yields a companion paradox, an updated version of Zeno's "argument against plurality." We are told that the shortest line segment contains an infinity of points. Then even the shell of a walnut can embrace a spatial infinity as imponderable as intergalactic space.

"Solid" matter is made of atoms—which are mostly empty space. The part that *isn't* empty space is protons, neutrons, and electrons. But these particles are mostly empty space too. If space is infinitely divisible, there may be an endless hierarchy of particles, subparticles, and sub-subparticles—all of which are *mostly empty space.* Then everything would be 99.999999+ percent nothing. It should be impossible to see anything—like Gertrude Stein's Oakland, there's no there there.

Physics provides a quick resolution for this paradox. Visible light scatters off electrons in atoms, which can act like waves extended in space. The electrons "smear out" and cover atoms, in effect. The fact that electrons can also act like infinitely small particles never enters into it. Nor does the nucleus of protons and neutrons play any role in scattering ordinary light.

For the paradox to work, you would have to have some kind of magic X-ray vision that allows you to see something if and only if a geometrically perfect line connects a point occupied by matter to your eye. Then you would not see a walnut, but the myriad point-like electrons and quarks composing it (or the ultimate subparticles of which electrons and quarks are made). Everything would be a fractal dust. Since you can't see a single infinitely small point, everything should be invisible.

Olbers's Paradox Resolved

That leaves Olbers's macroscopic paradox. Any resolution must lie in the premises: that the universe is infinite, that stars are scattered randomly, that nothing prevents the light of distant stars from reaching us. All three explanations have been advanced.

One approach is to suppose that the distribution of stars is something like the distribution of subatomic matter in the discussion above. The two complementary paradoxes annihilate each other. C. V. L. Charlier, a Swedish mathematician, resolved Olbers's paradox by proposing that stars are not scattered haphazardly but cluster in ever-greater hierarchies. We now know that all the nearby stars are part of a galaxy, the Milky Way, and the Milky Way is itself part of a cluster of galaxies called the Local Group. The Local Group is part of a yet greater hierarchy called the Local Supercluster. The Local Supercluster is part of the Pisces-Cetus Supercluster Complex . . . If and when someone announces that the Pisces-Cetus Supercluster Complex is part of something even bigger, no one will be much surprised.

Charlier showed that, under an endless chain of hierarchies, paradox can be avoided even if the number of stars is infinite. You might have, for instance, a super-super-supercluster of galaxies so remote that its image in our sky could fit behind the tiny orb of Arcturus or Betelgeuse. There would be super-super-super-superclusters and super-super-super-super-superclusters so much farther away that they appear smaller yet. Under Charlier's scheme, you could travel endlessly in most directions and never come to a star. Thus the night sky is dark.

Charlier's explanation is geometrically possible. It fails only in that it does not seem to describe the relative distances and dimensions of the observed cosmological hierarchies. Nearby galaxies loom much larger than nearby stars. Though extremely faint, the Andromeda galaxy is several times the apparent diameter of the sun or the full moon. The Magellanic Clouds of the southern sky (the two galaxies nearest our own) are about the size of lemons held at arm's length. And nearby clusters of galaxies are bigger yet. The Virgo Cluster, invisible to the eye, sprawls over an entire constellation.

Current thought on Olbers's paradox invokes a fact not suspected until this century: The universe is expanding. All the distant galaxies we can see are receding from our own galaxy at great speed. We cannot measure this motion directly, of course, but it produces a telltale shift in the light we receive from the galaxies, and attempts to explain this shift in any other way have failed. The galaxies in any given part of the sky are moving away from us; and the galaxies in the opposite part of the sky are also moving away from us *in the opposite direction.*

One interpretation of this is that our galaxy is "special," the center of the universe. The facts are equally well explained by postulating that the whole universe is expanding. That way of putting it is convenient but a bit misleading. It is not a uniform expansion like the one Poincaré talked about, but an expansion at the very top of the scale of distances. Neither the earth nor the Milky Way is getting bigger—and perhaps not even the Local Group. But the distances between galactic clusters *are* getting bigger. In principle, we can measure the ever-widening intergalactic gulfs with our yardsticks, which have not expanded.

Under the hypothesis of universal expansion, there is nothing unique about our galaxy or its place in the universe. The inhabitants of those distant galaxies would find themselves the "center" of expansion too. Since this hypothesis does not demand the extraneous assumption that our galaxy is special, it is favored.

The most distant galaxies known are hurtling away from the earth at near the speed of light. The light given off from a rapidly receding object undergoes a "red shift." This increases the light's wavelength and reduces its energy. Energetic visible light is redshifted to low-energy microwaves. When a luminous object is moving away at nearly the speed of light, the energy is attenuated almost to the vanishing point. Thus the light we receive from very distant galaxies is of such low energy as to be invisible.

Let's see how this affects Olbers's reasoning. Think of the universe as being divided into a series of concentric "shells" of space centered on the earth. Because light attenuates with the square of the distance, the amount of light we receive from each shell should (on the average) be equal. All the stars within 10 light-years of the solar system should generate about as much light as the stars between 10 and 20 light-years, or between 30 and 40 light-years, or for that matter, between 1,000,000 and 1,000,010 light-years.

If the universe is infinite, the total amount of light we receive is the sum of an infinite series: something like $x + x + x + x + x + $

. . . , where x is the light from each shell. This type of infinite series does not converge but rather adds up to infinity.

When the light from the more distant shells is weakened by the red shift, it changes everything. The more distant the galaxy, the faster it recedes, and the less energetic its light. The infinite series might look more like this: $x + 0.9x + 0.81x + 0.729x + 0.6561x + \ldots$ An infinite series like this, where each term is smaller by a fixed factor, does converge. An infinity of stars could shine in the earth's sky and still produce only a finite amount of light.

Few cosmologists doubt that the expansion of the universe is an acceptable explanation of the paradox, but there is a simpler explanation. In 1720 Edmund Halley wrote that the darkness of the sky argued against an infinity of stars. Today, many cosmologists believe the universe *is* finite (though for reasons other than Olbers's paradox). General relativity provides a way for the universe to be finite without ever coming to a worrisome "end." Space could curve back on itself, a three-dimensional analogue of the surface of a sphere. If you could walk far enough in any direction on the earth, you would come back to where you started. Space itself might be like that: A rocket traveling far enough in a straight line would return to its point of launch.

Most current cosmological models predict just such a finite universe, provided the density of matter in the universe equals or exceeds a certain limit. The observed density of visible matter (stars) is below the limit, but it is conjectured that there is enough invisible matter (intergalactic hydrogen, black holes, neutrinos?) to create a finite universe. Recent studies of "gravitational lens" effects of distant galaxies and quasars support the belief that there is much invisible matter.

The Paradox of Tristram Shandy

There is a subconscious double standard: Infinities of time seem a little different from infinities of space. It is natural to think that space extends out in all directions forever (or is this a culturally instilled belief?). Time is supposed to be infinite *only* in the future direction. We ask when time began but rarely where space began.

The infinity of past time is an unpopular belief. Yet it would "answer" questions of when or how the world was created by throwing them out as meaningless. In contrast, the infinity of future time appears to be universally accepted, even by those religions that postulate an apocalypse. After the millennium, the good live on forever, or the cycle starts over with a new creation. Few if any doctrines are nihilistic enough to believe in a *real* end to time, where things revert to exactly the same nonexistence as before the beginning of time, only this time for good.

Bertrand Russell's "paradox of Tristram Shandy" plays with the idea of an infinite future. Tristram Shandy is the raconteur narrator of Laurence Sterne's rambling novel of the 1760s, *The Life and Opinions of Tristram Shandy, Gentleman*. Russell wrote: "Tristram Shandy, as we know, took two years writing the history of the first two days of his life, and lamented that, at this rate, material would accumulate faster than he could deal with it, so that he could never come to an end. Now I maintain that, if he had lived for ever, and not wearied of his task, then, even if his life had continued as eventfully as it began, no part of his biography would have remained unwritten."

Russell's reasoning goes like this: Say that Shandy was born on January 1, 1700, and began writing on January 1, 1720. The first year of writing, 1720, covers that first day, January 1, 1700. The progress would go like this:

Year of Writing	Covers Events of
1720	January 1, 1700
1721	January 2, 1700
1722	January 3, 1700
1723	January 4, 1700
•	•
•	•
•	•
etc.	etc.

There is a year for every day, and a day for every year. Were Shandy still writing today, in 1988, he would be up to the events of September 1700. In turn, this immortal Shandy would get to setting down *today's* events circa the year 106,840. You cannot single out a day for which it *isn't* possible to schedule a future year for recording its events. Therefore, said Russell, "no part of his biography would have remained unwritten." Even so, Shandy gets further and further behind in his writing. With every year of writing he falls 364 years further from completion!

Russell's reasoning was based on Georg Cantor's theory of infinite numbers. If two infinite quantities can be placed in one-to-one correspondence to each other, they are equal. For instance, mathematicians hold that the number of whole numbers (0, 1, 2, 3, 4, 5

. . .) equals the number of even numbers (0, 2, 4, 6, 8, 10 . . .)—
rather than being twice as much, as you might think. The two are
equal because every whole number n can be paired with one even
number $2n$, and the pairing will leave no even numbers left over.

More mind-bending is a reversal of the paradox discussed by
W. L. Craig. Suppose that there is an infinity of *past* time, and that
Shandy has already been writing for an eternity. Then, suggested
Craig, there is the same Cantorian correspondence between years
and days. Shandy would have just finished the last page of his auto-
biography. But that's ridiculous. How could Shandy have chronicled yesterday's events already, when it should have taken him a
whole year?

Craig and others have used the reverse paradox to demonstrate,
not very convincingly, the impossibility of a past eternity. A reasonable resolution of the reverse Tristram Shandy paradox was supplied by Robin Small. It is, in fact, impossible to establish a correspondence between *specific* days and *specific* years.

Pretend it is midnight, December 31, 1988, and Shandy has just
finished the last page of his manuscript. What day has Shandy been
writing about this past year? It can't have been any day of this year.
(Otherwise he would have spent the early part of the year writing
about a day that hadn't happened yet.) The most recent day he
could have been writing about in 1988 is December 31, 1987.

If indeed Shandy spent 1988 recounting December 31, 1987, then
he must have spent 1987 writing about December 30, 1987. That
again is impossible. Actually, Shandy couldn't have written about
any day later than December 31, 1986, in 1987.

But if he wrote about December 31, 1986, in 1987, he would have
had to write about December 30, 1986, in 1986 . . . Any proposed
correspondence crumbles beneath our feet. The alleged day Shandy
has been writing about recedes into the infinite past. There is no
way of singling out any day.

Conclusion: If there is an eternity of past time, and Shandy has
been writing since the beginning, he will have an infinitely long
unfinished manuscript. The most recently completed page will describe events of the infinitely remote past.

Russell's and Craig's versions of the paradox are not so different
after all. Russell does not claim that Shandy will "ever" finish the
manuscript. Rather, no specific day you can mention goes unre-
corded. Tristram Shandy's "last" page is forever a mirage.

9

NP-COMPLETENESS

The Labyrinth
of Ts'ui Pên

JORGE LUIS BORGES'S STORY "The Garden of Forking
Paths" describes a labyrinth so intricate that none escape from it.
Upon receiving road directions, the narrator digresses:

> The instructions to turn always to the left reminded me that such was
> the common procedure for discovering the central point of certain laby-
> rinths. I have some understanding of labyrinths: not for nothing am I
> the great-grandson of that Ts'ui Pên who was governor of Yunnan and
> who renounced worldly power in order to write a novel that might be
> even more populous than the *Hung Lu Meng* and to construct a laby-
> rinth in which all men would become lost. Thirteen years he dedicated
> to these heterogeneous tasks, but the hand of a stranger murdered him
> —and his novel was incoherent and no one found the labyrinth. Be-
> neath English trees I meditated on that lost maze: I imagined it invio-

late and perfect at the secret crest of a mountain; I imagined it erased by rice fields or beneath the water; I imagined it infinite, no longer composed of octagonal kiosks and returning paths, but of rivers and provinces and kingdoms . . . I thought of a labyrinth of labyrinths, of one sinuous spreading labyrinth that would encompass the past and the future and in some way involve the stars.

The term "labyrinth" is of uncertain and very ancient origin. In classical times, a labyrinth was a building, at least partly underground, of intentionally confusing design. Herodotus rated the Egyptian labyrinth near Crocodilopolis (completed in 1795 B.C.) a greater wonder than the pyramids. It contained 3000 chambers, half above and half below ground level. A forest of pillars stretched as far as the eye could see. Herodotus toured the upper half but was not permitted to descend below; there, he was told, sacred crocodiles guarded the tombs of kings. The progressive decay of this labyrinth is chronicled by a number of ancient writers, and its site was never lost. The foundation, unearthed in 1888, measures 800 by 1000 feet.

In Western tradition, the most famous maze is the Minotaur's labyrinth on the Greek island of Crete. In Greek legend, the Minotaur, a monster with human body and bull's head, inhabited the center of a vast maze designed by Daedalus for the Cretan king Minos. After Crete's defeat of Athens, Minos decreed that the citizens of Athens sacrifice seven young men and seven young women to the Minotaur every nine years. None of the youths who entered the Minotaur's labyrinth ever made their way out. The Athenian prince Theseus volunteered for the sacrifice. Minos's daughter, Ariadne, advised him to let out a silken thread as he entered the maze so that he would be able to find his way out. In this way Theseus slew the Minotaur and ended the tribute.

The legend may have evolved from travelers' tales of Athenians sent to Crete to pay tribute during the height of Minoan sea power. Maybe they saw something in Crete they didn't understand (a mystery-cult priest wearing a bull mask?), and the story became garbled. It is unknown whether, or in what form, a labyrinth existed in ancient Crete. In the Cretan language, a labyrinth could mean a mazelike building, a grotto or winding cave (a common feature of the Cretan landscape), or an inescapable dilemma in argument: a paradox. After Cretan civilization ebbed, the ruined palace at Knossos was called a labyrinth, perhaps only in sardonic comparison to a rock cave. Surviving coins of Knossos show the plan of a maze that seems to be an architectural labyrinth, not just a natural cave. Archaeologists discovered the remains of the palace at Knossos early in the twentieth century, but nothing resembling a labyrinth was found.

Another maze shrouded in legend is Rosamond's Bower, supposedly built in a park at Woodstock, England, in the twelfth century. The impenetrable maze concealed King Henry II's mistress, Rosamond the Fair (c. 1140–c. 1176), from his wife, Eleanor of Aquitaine. Henry found his way to a hidden trysting place aided by a secret "key" that revealed the route. According to legend, Eleanor found her way to the center of the maze trailing string and made Rosamond drink poison. This story did not appear until the fourteenth century and is apocryphal. It is not even certain that Rosamond's Bower existed, or if it was a proper maze in the modern sense. Less romantic historians suspect it was only a poorly designed house with confusing passages.

The modern "labyrinth" is almost always a garden maze of paths bounded by hedges of (in Britain) hornbeam or yew. British garden mazes flourished during Tudor and Stuart times. Maze designers often incorporated a key, a clandestinely marked route of egress, so that the initiated could find their way in and out without difficulty.

The labyrinth remains mysterious. The problem of threading a maze is NP-complete: one of a set of universal problems with the potential to confound the most powerful of computers.

NP-Complete

The world is a labyrinth of madly interlocking connections and relationships. One idea that expresses this goes by the blandly understated name "NP-complete." For the record, "NP-complete" stands for "nondeterministic, polynomial-time–complete." Those daunting words name a fundamental and universal kind of problem —one rich in practical and philosophical significance.

NP-complete is a class of problems that have been haunting computer programmers for three decades. Computers have been getting faster and more powerful ever since their invention. The computers of the late 1980s are roughly 30,000 times faster than the computers of the late 1950s. One boast goes: "If automobile technology had advanced at the same rate as computer technology, a Rolls-Royce would travel at supersonic speed and cost less than a dollar." By the mid-1960s, however, computer scientists began to realize that something was amiss. Certain common problems are extremely difficult to solve by computer (or by any known method). Throwing faster processors or more memory at them doesn't make nearly as much of a difference as might be hoped. These problems came to be called "intractable" or "intrinsically difficult."

One example is the "traveling salesman" problem, which appears in many old puzzle books. A mathematical riddle, it asks you to find the shortest route for a salesman who must travel to several cities, given the distances between the cities. The problem taxes the power of the largest computers. The catch is combinatorics, the stupendous number of combinations of a not so large set. The best ways known of solving the problem are not that much faster than adding up the mileage for every possible routing. The amount of arithmetic to be done mushrooms as the number of cities increases, and quickly outstrips the capacity of any conceivable computer.

The class of NP-complete problems was first explicitly described in a 1972 paper by Richard M. Karp of the University of California at Berkeley. Since then NP-complete has reared its head in dozens of unexpected areas. Many children's riddles, puzzles, games, and brainteasers are instances of NP-complete problems. It almost seems that a short problem has to be NP-hard to provide much of a challenge. Research into NP-complete problems has often been vigorously funded, by the standards of theoretical research, because of the enormous economic value. Virtually any industry that does computer modeling, from oil production to integrated circuit design, faces NP-complete problems. Finding an efficient "solution" to the NP-complete problems (deemed unlikely by most computer scientists) would be worth many billions of dollars. There is no more poignant illustration of the information age than that much of the military security of the United States, the Soviet Union, and other technologically sophisticated nations is now perched precariously on NP-complete. The "public key" ciphers that protect the superpowers' sensitive data are founded on the practical insolubility of large NP-complete problems. Similar ciphers promise to ensure privacy of personal data in business and governmental data bases. The discovery of the equivalence of so many diverse problems is intriguing from a philosophical viewpoint too. It is little wonder that "few technical terms have gained such rapid notoriety as the appellation 'NP-complete,' " as Michael R. Garey and David S. Johnson began their influential 1979 book *Computers and Intractability: A Guide to the Theory of NP-Completeness.*

NP-completeness is such a slippery abstraction that it is well to describe it with a concrete symbol. A labyrinth is more than a metaphor for our quest for knowledge; it is (from a suitably abstracted perspective) methodologically equivalent to our logic. Mazes prefigure the essential problem of deduction, that of finding a paradox.

Maze Algorithms

Let's approach NP-completeness by asking this: Is there a general method that will solve any maze?

Yes; there are several methods, in fact. Not all mazes are puzzles. A *unicursal* maze is one with a single unbranching path from beginning to end. You can't make a wrong turn. Many medieval mazes were convoluted but unbranching paths leading to a tree or a shrine. Early British churchyard mazes of this type symbolized the tortuous path of the pious through the wickedness of the world. Pilgrims traversed some mazes on their knees. At each turn of the maze, the pilgrim might have said *pater* or *ave*.

It is possible that the Minotaur's labyrinth at Knossus was unbranching. The coin design shows an unbranching path. If you met the Minotaur in such a maze, you would have only to do an about-face and run; you could never find yourself backed into a blind alley. On the other hand, the coin may show a stylized motif and not a literal map; or the design may have represented the correct route to take in a more complex network of paths. Using a silk thread to find the way out makes no sense unless there were branching paths.

Every maze has at least one entrance, and most have a goal, a point in the maze you try to find. The goal is usually near the maze's center, though it may be simply an exit on the maze's perimeter (as in amusement-park halls of mirrors). To solve such a maze is to find a route from entrance to goal. There may be just one route, or many. When more than one route connects the entrance and goal, the implicit puzzle is to find the shortest route.

Some mazes have more than one entrance. They are not fundamentally different from single-entrance mazes. There your first choice is which entrance to use; the fact that this choice is made outside the maze walls does not really change things. There are also multiple-goal mazes where the visitor is expected to visit all parts of the maze, or all of a set of points marked with statuary, benches, or other devices. Louis XIV built a celebrated labyrinth at Versailles in which visitors sought out each of thirty-nine sculptures commemorating Aesop's fables. A water fountain sprang from the mouth of each animal that spoke in the fables. Finally, there are undirected mazes where the point is simply to go in, wander around, and find your way out again.

In the geography of mazes, a *node* is a fork, a point where paths meet and you must make a decision. The entrance, the goal, and dead ends are also considered nodes. The segment of path between two nodes is called a *branch*. A simple map of any maze can show the nodes as dots linked by lines representing the branches. A maze is a graph, in the mathematical sense of the term.

Nearly all physical labyrinths are two-dimensional. That means branches never cross. In a three-dimensional labyrinth, bridges and underpasses permit branches to cross over each other like freeway interchanges.

It is one thing to solve a maze from a diagram and another to solve it from within. The eye can often solve the paper mazes of puzzle books at a gestalt. Inside a real maze of hedges or masonry it is difficult to form a mental map. Clever designers may make one junction of branches look so similar to another that you think you're going in circles even when you aren't. Nor can you resort to such time-honored paper techniques as starting at the finish and working backward (sometimes this is easier, sometimes not) or marking off dead ends to make the through routes more visible.

The difficulty of a maze has a lot to do with the number of branches leading from each node. When each node is allowed only one branch, the only possible arrangement is a unicursal maze. Represent the nodes as two beads, each connected to one end of a string. No matter how you convolute the string, it leads inexorably from one bead to the other. The maze at Chartres Cathedral is unicursal. It has no walls but is laid out in blue and white marble on the floor.

Nor is there a puzzle when two branches meet at each node. Picture a length of string with a number of beads along its length. There is still no choice to make. In fact, normally you don't even count a "junction" of two branches as a node. It is simpler to think of it as a kink in a single branch.

A real fork in the maze requires at least three paths meeting at a point (the fork's "handle" and two "tines"). The more branches meeting at a maze's nodes, the more difficult it is.

By custom, most recent garden labyrinths have an approximately rectilinear design in which substantially all the area is honeycombed with paths and hedge dividers. This makes it difficult for more than four branches to meet at one node. A more flexible design was used in the labyrinth at Versailles. There branches were not necessarily parallel, and many might meet at a single node. The maximum was five branches at a node. André Le Nôtre, architect of the Versailles maze, built another labyrinth at Chantilly containing a central node where eight branches joined.

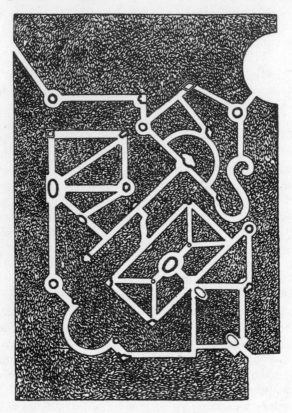

The Versailles Labyrinth

The table shows some statistics on several famous labyrinths. In some cases, the number of nodes and branches is open to interpretation. In each case I have tried to count as a node each point where a careful traveler would come to a decision. I counted entrances, goals, and dead ends as nodes, but did not count any superfluous nodes where two branches met. The last column, the average number of branches meeting at a node, is a rough measure of the difficulty of the maze.

Labyrinth	Nodes	Branches	BRANCHES PER NODE	
			Maximum	Average
LA LIEUE Chartres Cathedral Chartres, France (built 1220)	2	1	1	1.00
HAMPTON COURT MAZE Kingston, England (1690)	16	16	3	2.00
CHEVENING MAZE Chevening, England (c. 1820)	18	22	4	2.44
VERSAILLES MAZE Versailles, France (1672; destroyed 1775)	30	43	5	2.87
LEEDS CASTLE MAZE Maidstone, England (1988)	27	41	4	3.22

The Right-Hand Rule

The best-known maze algorithm is the "right-hand rule." Whenever you have a choice, take the rightmost branch. If you come to a dead end, retrace your steps to the last node and take the rightmost of any branches not yet visited. The best way to visualize these rules is to keep your right hand touching the hedge to your right throughout the maze. Never skip over a branch on your right.

Of course, there's nothing special about the right hand. The "left-hand rule" works just as well. It is necessary only that you be consistent once you enter the maze.

Why does this work? It is more universal than a simple convention such as "turn a screw clockwise to tighten." You could make a screw that turns the other way. The right-hand rule follows from the topology of the maze.

Think of a plan of a maze on paper. The regions of hedge are colored green. The white area between the hedge regions is the navigable part of the maze. In many mazes, the hedge region is all in one piece. There is just one hedge, convoluted as it may be. It looks like a strangely shaped country. Like any country on a map, the green region has a boundary. This boundary (which corresponds to the wall of the maze) is a single closed curve. Any part of this curve is connected to any other part. Therefore, patiently keeping a hand on the hedge wall will lead you to all parts of the maze.

This has virtually the same rationale as the Boy Scouts' algorithm for finding your way to civilization by following a stream. All of North America is one continent; the coast of North America, including the indentations of rivers and streams, is a closed curve. Following riverbanks or coastline must eventually lead to New Orleans (or any river/coastal point). A maze may have some disconnected islands of hedge, but as long as the hedge around the entrance and the goal is part of the same island, the rule works.

The right- or left-hand rule has the virtue of simplicity. It has two defects. First, it is inefficient. It sends you down every cul-de-sac on the right (left) side. Most of the time, there is a much shorter route to the goal. Worse yet, the right-hand rule does not work for all mazes. It apparently would have worked for all known garden mazes constructed up until the 1820s. Then the Earl of Stanhope, a mathematician, devised a more difficult maze that was planted at Chevening, Kent.

The Chevening maze defeats the right-hand rule through eight separate "islands" of hedge. The entrance and goal are not on the same island. Applying the right-hand rule will take you around one region, but you will never see the goal. (Following riverbanks and coastline on Long Island will never get you to New Orleans.) Such a maze demands a more sophisticated algorithm.

The Trémaux Algorithm

All the more powerful ways of solving mazes require some way of making sure you aren't going in circles. You need to mark your path by letting out string, dropping bread crumbs, bending twigs, or the like. Or else you must have a *very* good memory for shrubbery, and be certain you can recognize all points previously traversed.

One general method, sure to solve any maze at all, is called the Trémaux algorithm, after an M. Trémaux mentioned as its inventor in French mathematician Edouard Lucas's *Récréations mathématiques* (1882). It can be thought of as an elaboration of Theseus' method of letting out string.

Theseus' string ensured that he could retrace his steps to the entrance without becoming lost. The string did not lead him to the

Chevening Maze (outer "island" of hedge shown in darker gray)

Minotaur's lair. Theseus might come to a fork in the maze and see that he has been going around in circles. It is reasonable that he could use this information to better decide which branch to take next. The Trémaux algorithm does just that.

Enter the maze. At first, go any way you like, marking your trail with string or whatever is handy. Continue until you come to the goal (if you are lucky), a dead end, or a fork in the maze you have visited before (as evidenced by your trail marks).

If and when you come to a dead end, backtrack to the previous node. (You have no alternative!) Be sure to mark even your back-tracking. Once you have entered and backtracked from a cul-de-sac, it will show two trails of bread crumbs. That tells you to avoid it in the future. In the Trémaux algorithm, you *never* traverse any branch more than twice.

When you come to a node in the maze you have visited previously (even via a different branch), do this:

If you approached on a fresh branch (only one bread crumb trail behind you), backtrack on that same branch to the previous node. Otherwise:

If there is an untrod branch leading from the node, take it. Otherwise:

Take any branch that has been traveled only once.

These are the only rules needed. Following the Trémaux algorithm will take you on a complete tour of the maze, in which every branch is traversed twice, once in each direction. Actually, you can stop when you find the goal; you do not have to travel the full circuit.

Like the right-hand rule, the Trémaux algorithm is extremely inefficient. Although you might luck out and take the most direct route from entrance to goal, chances are that you will traverse much or all of the maze before finding the goal.

It's never "too late" to use the right-hand or the Trémaux method. You can enter a maze, go any which way you please, and resort to algorithms only if you get lost. Think of the arbitrary point at which you start implementing the algorithm as an alternate "entrance." The Trémaux algorithm will take you on a complete tour of the maze starting at that point, including the goal and the real entrance. Both methods work in confusing buildings as well as garden mazes. If you were lost in the Pentagon or the Louvre, you could use the Trémaux algorithm to find an exit.

An Infinite Labyrinth

Imagine a labyrinth of infinite extent. Since the labyrinth is endless and covers the whole world, there is no entrance, no perimeter at all. You start your explorations from an arbitrary point in the labyrinth, no more knowing where you are in the grand scheme of things than we know where our galaxy is in the universe.

The design of the labyrinth is simple. At every node, exactly three branches meet. Nodes are distinguished by landmarks—assorted statues, benches, trees, and so on—that you may search for.

Like all labyrinths, the prime characteristic of this one is its irrationality. In looking for a given goal, there is no basis for preferring one path over another. Any path *could* be "right." It depends on how the maze was built. There is a vertiginous feeling in knowing that the labyrinth repeats itself—with variations—endlessly. Let a traveler spend years exploring a certain region of the labyrinth and come to a fork on the frontier of the known region. One of the unexplored branches leads to a desired landmark: Which one? The fact is, the explored portion of the labyrinth must be exactly repeated many times in the labyrinth. In some of the repetitions, the familiar paths are connected to the rest of the maze so that the right branch leads to the landmark; in others, the left branch does. The traveler, of course, has no way of knowing which applies in his case. This makes a mockery of any attempt to rationalize which path to take.

Suppose you find yourself in this infinite maze, wander for a time, and become hopelessly lost. You have not marked your trail and aren't sure exactly how far you have come.

You wouldn't want to use the Trémaux algorithm in that predicament. The Trémaux method doesn't constrain your movements until you start crossing your own path. You could go miles deeper into the labyrinth, getting more and more lost. It's even possible, in an infinite labyrinth, that you would never cross your own path, never see the goal, and never see any familiar point again.

Both the Trémaux and the right-hand method presume that it wouldn't be too bad to traverse the whole maze or a big part of it, just as long as you eventually reach your goal and don't keep going in circles endlessly. The Trémaux method actually favors exploring distant regions of the labyrinth first. You always take an untrod branch over a familiar branch, and avoid crossing your own path until absolutely necessary. In a finite garden maze, this is sound advice because the goal is almost always relatively distant from the entrance. There is little point in using up real estate and paying hedge trimmers' salaries for a maze that is larger than its puzzle demands.

In an infinite maze, you can't afford to wander aimlessly through parts unknown. When you are lost but know that a goal is relatively close (compared to the overall dimensions of the maze), you should explore the nearby regions first, progressing outward as necessary. An algorithm that does that was described by Oystein Ore of Yale in 1959.

The Ore Algorithm

The scheme is easiest to explain if you start from a node. If you aren't at a node, go to the nearest node. If you don't know which direction leads to the nearest node, go in either direction to the first node. Then mark that node somehow. It will be your home base.

Starting at the home node, explore each branch leading from it. Place a pebble at the entrance of each branch as you enter it. Explore each path only to the next node. Then place a pebble at the far end of the path and retrace your steps to home base.

Identify any dead ends. Once so identified, a branch can be ignored in the future. Mark dead ends by blocking them off with string or putting a line of pebbles across the entrance.

Should a path loop directly back to the original node, mark it as a dead end too. It is equally profitless.

You are interested in locating those branches that lead to new nodes with new branches. At the end of the first stage of exploration, each potential route to the goal has a pebble at either end, and you are once again at the home node.

Next, explore to a depth of two nodes. Travel up each non-dead end branch to the new node, and explore each of the branches radiating from it in the same manner. Add a pebble to each end of the primary branches (so they now have two pebbles at either end), and put a pebble at either end of the new secondary branches. This prevents you from being unable to find your way back to the home node; the branch leading to it has one more pebble than the others. As before, mark off entrances to dead ends or loops. If a branch leads to a previously explored node (one with at least one marker pebble), mark that path off at both ends too.

At the third stage of exploration, travel three nodes deep from the home node, adding a pebble to each end of each explored branch. Continue exploring further and further out until the goal, entrance, or other desideratum is found.

The Ore algorithm will locate the shortest route to the goal (measured in branches, not distance as the crow flies). The course of your explorations won't be this shortest path, of course, but if the shortest path traverses five nodes, you will find it in the fifth stage of exploration and know that path to be a minimum.

The Ore algorithm is woefully inefficient too. Rather than automatically zeroing in on the right route to a goal, it checks all possible routes. It *has* to because any route could be the right one.

NP-Completeness of the Maze

Consider what might be called the eternal question of the labyrinth. You are at a point E (for "entrance," although, like all other

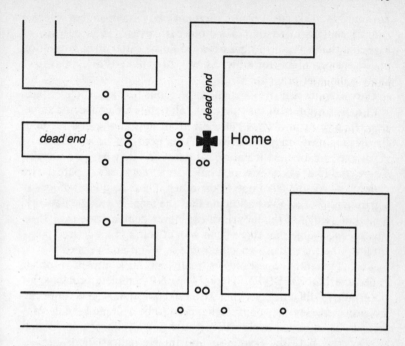

points, it is lost in the unboundedness of the infinite labyrinth). You seek a point G, a goal that is also an arbitrary point in the maze. You do not know where G is, in the sense of being able to locate it on a map (of which there are none). You are sure you can recognize your goal—if and when you come to it—by a landmark known to exist at point G. The ever-present question, posed implicitly by the very fact of the maze, is: "What simple route(s) connects E to G?"

A *simple* route is one that doesn't cross itself—one where you don't find yourself going in circles. It's never necessary to go in circles, so a simple route is of primary interest. There may be more than one simple route. If so, you prefer the shortest route, but you do not worry too much about such niceties. Faced with the formidable problem of exploring the infinite maze, you would be happy with almost any route to point G.

Closely related to the eternal question of the labyrinth is an easier question that might be called the existential question of the labyrinth. It asks, *"Is there* a simple route from E to G?"

Note why this is easier. Whenever it is possible to answer the eternal question (by specifying a route), it is simplicity itself to answer the existential question: The answer is yes. Even when it is not possible to specify a route, there may be circumstances where it can be demonstrated that a simple path exists. It is only to be expected that a yes-or-no question would be easier than a question whose answer might (for all we know) entail tedious directions for a route billions of branches long.

Only skeptics ask the existential question. It is an article of faith with most explorers of the maze that all points are connected somehow, that you can always get there from here. Long and tortuous though the route may be, it exists. This need not be so. The labyrinth may not be fair; it may pose questions that have no answer. There could be two intertwined but distinct networks of paths, with no way of crossing from one to the other. There could be trillions of separate networks. Even allowing that the maze is a single network, local knowledge of the labyrinth can never demonstrate this. It remains conceivable that there is no way of getting to a desired point, up until the time that a specific route is found and verified.

The "existential question" is really an NP-complete problem called the LONGEST PATH problem. NP-complete problems are notoriously "difficult," yet the existential question is sometimes easily answered. Say that point G happens to be a single branch away from E. Then casual exploration would find G almost immediately —answering both the existential and the eternal question.

Nothing wrong there. Specific instances of a general problem may be quite easy. What is sought is a general, systematic method of answering the existential question that would work in the smallest maze or in an infinite labyrinth.

There is no fast way of solving an unknown maze, no way of knowing, presciently, which paths to favor. The best one can do is to examine nearly *all* paths until the goal is found. The various maze algorithms only prevent one from repeating the same branches unknowingly or from wasting further time with known cul-de-sacs and loops. The algorithms cannot guide you "intelligently" through untrod regions of the labyrinth.

Look at the Ore algorithm, as efficient as any. You start from the home node. Three branches lead from this node. Each neighbor node is connected to two other nodes. (One of the three branches is the one connecting the neighbor node to the home node, which has already been considered.) In turn, each of the six nodes once removed is connected to two more nodes. The maze is hydra-headed. Branches you explore lead to new nodes from which spring yet more branches. Some of these branches may have been explored previously (as evidenced by old trail markers). Much of the time, the number of branches to be explored mushrooms exponentially. The more you know about the labyrinth, the more you realize you don't know.

If you can explore a branch a minute, the progress of the Ore algorithm looks like this:

STAGE OF EXPLORATION	ROUTES	BRANCHES EXPLORED IN THIS STAGE	CUMULA-TIVE NUMBER OF BRANCHES	TIME REQUIRED
1	3	6	6	6 minutes
2	6	24	30	30 minutes
3	12	72	102	1.7 hours
4	24	192	294	4.9 hours
5	48	480	774	12.9 hours
10	1,536	30,720	55,302	38.4 days
15	49,152	1,474,560	2,752,518	5.23 years
20	1,572,864	62,914,560	119,537,670	227 years
30	1,610,612,736	9.66×10^{10}	1.87×10^{11}	355,000 years
45	5.28×10^{13}	4.75×10^{15}	9.29×10^{15}	17.7 billion years

In all finite labyrinths, the process of discovering new branches must eventually come to an end. Past a certain stage in the exploration, most new branches will lead back to familiar nodes. Finally, all the branches will have been traversed, and the goal must be known. In the infinite labyrinth, the exponential spiral continues forever. Even when the goal is relatively nearby it can take impracticably long to find it. It would take all day to find a goal five nodes away, though the route itself, once known, could be traversed in five minutes. The search for a goal fifteen nodes away would take years, and not all the time in the universe would be adequate to find a goal just forty-five nodes distant.

Look at the LONGEST PATH problem from the point of view of a computer programmer. You want a computer to decide whether there is a route connecting two points in a certain large labyrinth. To do this, you must give the computer a "map" of the maze. This map takes the form of a list of all the nodes in the maze and a list of all the branches. The nodes are numbered or named; the branches are specified by indicating which nodes they connect and the distance (an integer expressing the distance in whatever units desired) between them. One branch might be listed as "Node 16, Node 49: 72 feet." The distance is the actual length of path a wanderer would cover, not straight-line distance. The two nodes that are the entrance and the goal are specified as such.

There is a further element to the LONGEST PATH problem, a specified distance, *n*. The LONGEST PATH problem asks whether there is a direct path between entrance and goal *longer* than *n* units of distance. If you like, *n* can be arbitrarily small or zero. In that case the LONGEST PATH problem asks whether there is a path longer than zero length—that is, whether there is any path whatsoever—between entrance and goal.

Since the existential question is NP-complete, the more difficult eternal question is at least as hard as the NP-complete problems. If it is impracticably hard even to say if a route to G *exists,* then it is also impracticable to specify such a route.

The Oracle of the Maze

The NP-complete problems have the surprising property that their answers are easily verified. You meet an oracle who has the ability to divine the answer, instantly, to any question at all. Those who believe in the oracle's omniscience come to him with questions so difficult that no one else can solve them, and he answers them immediately.

Yet he is frustrated in his attempts to prove his power to *everyone*. There are skeptics. The oracle wants to prove his omniscience is genuine by showing that the answers he gives are correct. This is not always possible.

He receives two types of questions. The most common type comprises difficult questions that no one else can answer. Why is there evil? Does God exist? What is the googol-th digit after the decimal point in the decimal expansion of pi? It so happens that the oracle's answers to these questions are correct and accurate in all respects. But he is unable to prove these answers. As the skeptics sneer, he could pretty much give any answer to these problems, and no one would be the wiser. Even the relatively down-to-earth questions (such as about the googol-th digit of pi) may be so difficult that the most powerful computer in the world cannot verify his answers.

To prove his powers, the oracle must answer questions whose answers can be checked. He gets many of these questions too, some from skeptics trying to show him up. What is the capital of Kiribati? What's the square root of 622,521? Name the sisters in *Little Women.* Here's a sealed box: What's in it?

The oracle answers each of these questions correctly, and the poser knows that he's answered correctly. The poser knows that because he knew the answers all along. That's the trouble. These questions are too easy to prove the oracle's powers beyond all doubt. If the person asking the questions already knew the answers by ordinary means, then conceivably the oracle knew or found out the answer by ordinary means too. His clairvoyance could be an act, the skeptics say; he could be nothing more than a calculating prodigy, well versed in trivia, who employs the paltry deceptions of the stage mind reader for the rest.

The oracle loses either way. Answer a question that no one else possibly could, and he is accused of fabricating; answer a question whose answer is known or knowable, and he is accused of cheating. To prove his omniscience, he needs a third type of question—a *difficult* question whose answer, once stated, can be verified *easily*. Are there questions like that?

Questions about the infinite labyrinth qualify. Let skeptics pick two random points in the labyrinth and ask the oracle to specify a route between them. Anyone can easily assure himself that an answer is correct (or incorrect). All they have to do is follow the prescribed route and verify that they end up at the right point.

Wouldn't this be another "easy" question? It is necessary to make sure that the two chosen points are far enough apart so that no one knows a route connecting them, or even could find such a route by normal means. The relative inefficiency of even the Ore algorithm guarantees that such pairs of points are common. If the points are twenty nodes apart, it would take centuries to find a path by ordinary means. Then wouldn't it be another "hard" question? No, because it would only take about twenty minutes to verify the oracle's answer (traversing a branch a minute). A maze's solution is much, much simpler than the maze itself.

This third type of question is close in spirit to what complexity theorists mean by the "class NP."

P and NP

There is a distinction between a problem in general and *instances* of a problem. A jigsaw puzzle is a general type of problem; a specific jigsaw puzzle with 1500 pieces that fit together to make a picture of a Dutch windmill is one instance of the problem.

The theory of NP-completeness judges the difficulty of problems not by any particular instance of the problem, but rather by the way the difficulty grows as a function of the size of the problem. In the case of a jigsaw puzzle, the "size" of a puzzle is the number of pieces. The more pieces, the harder the puzzle is. How "hard" the puzzle is is best measured by the time it takes to solve it. This depends on how fast you work, of course, but it clearly has a lot to do with how many pieces must be compared against other pieces to find a match.

In a worst-case jigsaw puzzle—such as one of those novelty puzzles that are all one color—you have to compare pieces randomly to see if they fit. In the early part of the process, you end up comparing each piece against a large fraction of the other pieces. The total number of distinct matching operations is proportional to the *square* of the number of pieces. Therefore, the time requirement can be expressed as a polynomial function containing n^2, where n is the number of pieces.

This time requirement is relatively modest. In a maze, the time required to find the goal using the Ore algorithm is more like 2^n, where n is the number of nodes to the goal. When n is small, the difference between n^2 and 2^n is not great. As n increases, a chasm opens up between the polynomial and exponential functions. You can solve a jigsaw puzzle with 5000 pieces. You cannot solve a nontrivial maze requiring 5000 correct turns to find the goal.

The "easy" problems of complexity theory are those general problems that may be solved in polynomial time. These problems form the class P (for polynomial). Think of class P as a vast country somewhere, poorly mapped, yet with distinct boundaries. Every point is either in P or not, though our maps are so unreliable that it is not always possible to say which. Jigsaw puzzles are one point in class P; so are simple arithmetic problems.

There is another class of problems, NP, including all those problems whose *solutions* may be *verified* easily (in polynomial time). When a problem is easy, its solution is easy to check too. If nothing else, you can check it by solving the problem all over and making sure you get the same answer. So all easy problems (class P) are in the class of problems whose answers are easy to check (NP). NP also includes many problems not in P, such as solving mazes. P is thus a province of the larger country NP. If we were to draw a map, it would look like the diagram on the opposite page.

The outer rectangle represents all possible problems. The class NP does *not* include all problems. There are ultra-hard problems whose answers cannot even be checked easily. These are represented by the region of the rectangle outside of NP's circle.

Put this in the context of the oracle's questions. The first class of questions, "hard" questions that cannot be checked, are analogous to the class of problems outside of NP. The second type of questions

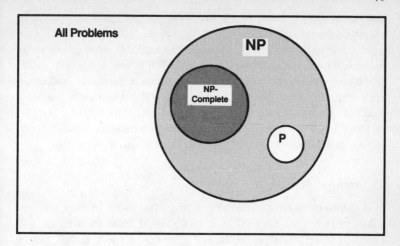

corresponds to the class P. The third class—hard questions with easily verified answers—corresponds to those problems in NP but outside of P.

The term NP ("nondeterministic, polynomial-time") refers to something known as a *nondeterministic* computer—specifically, the idealized computer known as a Turing machine, envisioned by computer pioneer Alan Turing. A nondeterministic computer is not quite what it may sound like. It sounds like a computer that works randomly or follows a less than exact "algorithm" (or a computer that has free will!).

The operation of a nondeterministic computer can be imagined this way: Instead of one computer we have a large and potentially infinite number of computers. Each computer is assigned one of the possible solutions of a problem, and is charged with checking out that solution.

If, for instance, the problem is to find a path through a maze, the ensemble of computers (actually robots in this case) starts at the entrance. Every time the army of robots comes to a fork in the labyrinth, they split up into as many parties as there are paths. The search parties keep splitting up at each new fork, and eventually all possible routes are explored.

At least one of the robots will indeed travel from entrance to goal. Let our attention focus on *that* machine. How long did it take? Chances are, it didn't take very long at all. Solutions to mazes are generally short; it is going down all the wrong paths and backtracking that makes them so difficult. The time a nondeterministic computer takes to "solve" a problem is exactly the time it takes to *check* a guessed solution.

The NP problems parallel the set of questions open to scientific inquiry. The scientist who tries to establish a new truth is in much the same position as the oracle above. Science is mostly concerned with hypotheses similar to answers to NP problems: hypotheses that may be verified or refuted readily.

There is an even more striking connection between the NP problems and science—logical deduction itself is an NP problem.

The Hardest Problem

What is the hardest problem in the class NP? In 1971 Stephen Cook proved that SATISFIABILITY is at least as hard as any problem in NP. His proof showed that no problem in NP can be any harder because all NP problems may be transformed into SATISFIABILITY problems.

This was followed (1972) by Richard Karp's discovery that many types of intractable problems share this distinction with SATISFIABILITY. Diverse problems from graph theory, logic, mathematical games, number theory, cryptography, and computer programming are equally as hard as SATISFIABILITY. The class of hardest NP problems is "NP-complete." In the Venn diagram, NP-complete is a circle in NP but outside P.

Strictly speaking, NP-complete is a shadow land that may not properly exist. It has not been proven that the NP-complete problems cannot be solved in polynomial time. The evidence is only empirical: For years, theorists and computer programmers have tried to come up with polynomial-time solutions to NP problems, and have always failed. In practice, proving that a problem is NP-complete is considered strong evidence that it cannot be solved efficiently.

It remains barely conceivable that every problem in NP can actually be solved in polynomial time by some unknown super-algorithm. In that case, P, NP, and NP-complete would be identical, and would be represented as a single fused circle.

If an efficient solution, a magic key, exists, then there are virtually no limits on what we may deduce, Sherlock Holmes-like, from logical premises. If, on the other hand, there is no efficient solution to SATISFIABILITY and the NP-complete problems, there is a world of truths that, as a practical matter, must go unrecognized. It is strongly suspected that there is no magic key: We are all Dr. Watsons who miss the implications of much of what we see.

This means that there is a relatively sharp cutoff in the size of a solvable logic problem. Just as a maze larger than a certain size will be practically unsolvable, so will a logic problem of greater than a certain complexity. It evidently follows that our deductions about the real world are limited too.

A Catalogue of Experience

Paradox is a more significant and far-reaching concept than it may seem. If one holds beliefs that are contradictory, then one cannot have justification for at least some of those beliefs. Without justification there is no knowledge. Therefore, understanding a set of beliefs entails (at a minimum) being able to detect a contradiction in those beliefs. For that reason, the problem of detecting paradox, SATISFIABILITY, is a delimiter of knowledge. The difficulty of SATISFIABILITY is inherited by any attempt to understand implications fully.

Newton's theory of universal gravitation was founded on nothing that the ancient Greeks didn't know. The germ theory of disease could have been advanced and confirmed centuries before it was, if someone had made the right connections. It follows that there must be yet undiscovered generalizations that are "overdue" right now. Quite possibly, we have all the necessary facts needed to deduce how to prevent cancer or the location of a tenth planet, but no one is putting them together in the right order. More than that: Maybe we're missing all sorts of logical conclusions about the world. They could be implicit in everything we see and hear, but might be just a little too complex to grasp.

"The grand aim of all science is to cover the greatest number of empirical facts by logical deduction from the smallest number of hypotheses or deductions," Einstein wrote. Take the sum of all human experience: everything that anyone ever saw, felt, heard, tasted, or smelled from the Ice Age up to this instant. This is the starting point for any codification of knowledge. In principle this information could be assembled into a vast catalogue. Let the catalogue be a simple list of experiences, without any interpretation of them. Dreams, delusions, hallucinations, mirages, and optical illusions are tabulated in detail, side by side with "real" experiences. It is left to the reader of the catalogue to determine which (if any) are the real experiences.

In the catalogue of experience must be every observation on which the naturalistic sciences are founded. A description of every bird, star, fern, crystal, and paramecium ever seen would be found somewhere. The catalogue would also contain the minutest details of every scientific experiment ever conducted. It would include both Michelson's and Morley's impressions of how the late afternoon sun glinted on their apparatus's mirrors on certain dates in 1887; the color, size, shape, velocity, and acceleration of every apple Newton ever saw fall.

Science is *not* simply a catalogue of experience. For one thing, no human mind can comprehend the totality of human experience. The catalogue must contain everything you've ever experienced—which has occupied 100 percent of your attention for your whole life up to now! Science compresses the human experience (certain aspects of it, anyway) into a manageable form. What we are really interested in is *understanding* the world described by the catalogue. That means seeing the generalities, if sometimes mercifully forgetting specifics. A vexing question in the philosophy of science is to what extent this is possible.

Every experience constrains the truth values of some of the unknowns of the world, as in a logic puzzle. The relationships between unknowns could be quite subtle, of course, and everything would have to be qualified with *ifs*. Perhaps one of your experiences is hearing your friend Fred tell about how he saw the Loch Ness monster last Tuesday. The actual import of this experience might be something like:

If Fred wasn't mistaken *and* Fred wasn't lying *and* the external world isn't an illusion, *then* the Loch Ness monster existed on Tuesday.

These *if* clauses are the inevitable auxiliary hypotheses that so complicate confirmation.

Feed the catalogue of experience into a super-computer, and program it to look for deductions. This task requires only logic, and logic is one thing computers do superbly well. When it is through, the computer could even sort the list of deductions according to importance, measured by how many distinct experiences they account for. The deduction at the top of the list would be the most important thing humanly knowable.

Fanciful as this idea is, it is a framework for introducing some of the deepest concerns of the philosophy of science. Sorites, the basis of our science, can be recognized and tested for consistency in polynomial time. These easy logic problems are comparable to unicursal mazes (one or two paths per node), which are trivial, unlike "regular" mazes, where at least some nodes have three or more paths. More complex deductions, involving premises with three or more unknowns, require (impractical) exponential time. There may be a whole world of logical deductions—of interpretations of our sense experience—that is forever concealed from us.

Think of our experience as a maze and of logical truths about that experience as paths through the maze. The NP-completeness of SATISFIABILITY suggests that we will never exhaust all the potential routes.

A Computer as Big as the Universe

Computer scientists Larry J. Stockmeyer and Albert R. Meyer dramatized the intractability of NP problems in a fantasy about a computer the size of the universe. They showed that the universe is not big enough for us to answer many questions about the universe.

Suppose we try to make a list of accepted beliefs. Like Descartes, we want to start with a blank slate and be very careful about adding beliefs to the list. Before any belief is added, it is first checked against the beliefs already on the list to make sure that it doesn't introduce a contradiction. This check is a SATISFIABILITY problem.

You might think that you could detect contradictions merely by running through the list and making sure that the proposed new belief does not directly contradict any belief already on the list. It's not that simple, though.

Granted, a new belief might contradict an old one. If the new belief is "All ravens are black," and one of the old ones is "No ravens are black," then you have a contradiction right there. Much more treacherous is the type of contradiction that arises from three or more statements that are tenable individually. Usually the term "paradox" is reserved for these latter cases, where the contradiction is not immediately evident.

Suppose the new belief is "All grass is green." The list could already contain this pair of statements:

All hay is brown.
Hay is grass.

Together with the new statement, this creates a contradiction. You might have missed it if you examined statements only in pairs:

Any two of the three offending beliefs are mutually compatible. To rule out cases like this, it is necessary to test each new belief with every other pair of statements in the list. This greatly multiplies the amount of checking to be done. And that still isn't the end of it. There may be subtler paradoxes that pop up only when sets of four, five, or more statements are considered together. Adding a belief to a set of a million can create a contradiction even when the new belief is compatible with every set of 999,999.

There is a lot of fact checking to be done, so clearly a computer is called for. We start with belief No. 1 *(Cogito ergo sum?)*. For the computer's benefit, the belief is encoded as a logical statement about Boolean unknowns. Next we get ready to add belief No. 2. We first instruct the computer to check it against belief No. 1 for possible contradiction. In this case there is only one logical test (belief No. 2 against belief No. 1).

Now the list has two beliefs and we want to add a third. The third belief must be checked three times: against No. 1, against No. 2, and against Nos. 1 and 2.

The fourth belief must be checked against *seven* sets: against Nos. 1, 2, and 3 collectively, against Nos. 1 and 2, against Nos. 1 and 3, against Nos. 2 and 3, and against each of the three beliefs individually.

In fact, a new belief must be checked against every possible subset of the current list. The formula for the number of subsets of n things is an exponential function: 2^n. This formula counts the null set, which we don't have to worry about. The number of nonempty subsets is $2^n - 1$.

Assume that the beliefs, or some of them, are logically complex enough so that there is no way of avoiding an exponential-time algorithm. Then the number of required comparisons can be gauged from this table:

SIZE OF LIST	SUBSETS
1	1
2	3
3	7
4	15
5	31
10	1,023
100	1.27×10^{30}
1,000	10^{301}
10,000	10^{3010}

Even a fairly modest list of beliefs—100, say—has an astronomical number of subsets. To qualify a 101st belief, it must be checked against over 10^{30} distinct subsets of the list.

How can that be? Isn't it obvious that you could write down 101 statements and *quickly* assure yourself that no paradox exists?

Indeed so. You could copy 100 random assertions from an encyclopedia, making sure that every sentence talks about something different. We are talking here of the more general case where the beliefs on the list will concern many of the same unknowns and be logically complex. The beliefs are allowed to intertwine like the premises in Carroll's pork-chop problem. Then we must fall back on an algorithm—a slow algorithm.

How fast could a computer add beliefs to the list?

Stockmeyer and Meyer's analysis had an "ideal" computer deciding the truth of certain mathematical statements with an exponential-time algorithm. Essentially the same reasoning applies to SATISFIABILITY problems. The power of any computer ultimately depends on the number of components it contains, Stockmeyer and Meyer said. The smaller the components, the more processing power can be packed into a given volume.

In the first digital computers, the logic gates were vacuum tubes and the connections were wires. Later, the vacuum tubes gave way to transistors. Currently, powerful processors fit on a single chip. Most interconnections are printed circuits of thin metallic film.

No one really knows how small a processor or a logic gate could be. Experimental gates use films that are only several atoms thick. There are promising technologies that have yet to be exploited. Stockmeyer and Meyer were wildly optimistic in their thought experiment. They postulated that, somehow, it is possible to construct computer components as small as protons. As it stands now, protons and neutrons are the ultimate in measurable smallness. So however small the components in an "ideal" computer might be, they cannot be any smaller than 10^{-15} meters across (negative exponents are fractions: that is 1 divided by 10^{15}, or a trillionth of a millimeter).

Assume that the proton-size components may be packed like sardines. Then any given volume can hold as many components as it could hold ideal spheres 10^{-15} meters in diameter. A computer the size of an ordinary personal computer (which has a volume of maybe a tenth of a cubic meter) could contain about 10^{44} distinct components. A minicomputer with a volume of a cubic meter would contain 10^{45}.

Another all-important factor in computer technology is speed. One bottleneck is the time it takes for a component such as a logic gate to switch from one state to another. The speed of light is the fastest that any form of information may be transmitted. At best, then, a component cannot switch faster than the time it takes light to cross it. If it did, one side of the component would "know" what was happening elsewhere faster than is permitted by relativity.

Light takes 3×10^{-24} seconds to cross the diameter of a proton. In Stockmeyer and Meyer's analysis, that was taken to be the switching speed of the components in the ideal computer.

In reality, computer speed also depends on how the components are interconnected and how well the available resources are marshaled for the problem at hand. Most present-day computers are serial, meaning that they do one thing at a time. At any instant, the computer is at one point in its algorithm. Much faster, potentially, are parallel-processing computers. Parallel computers contain many processors and split tasks among them. Most of the time, a parallel computer is doing many things at once.

Since we're not skimping, assume that the ideal computer implements an ultra-sophisticated parallel-processing scheme. Every proton-size component is a distinct processor, and all are linked in some Connection Machine-like scheme that ensures relatively direct connections even when the number of processors is astronomical.

The computer splits its task among its processors by assigning each a distinct subset of the current list of beliefs. Let each processor be able to compare a new belief against its present subset *instantly*. It can determine whether there is a contradiction and fetch a new subset to test in its switching speed of 3×10^{-24} seconds (call it an even 10^{-23} to simplify the math). Then each processor can run through 10^{23} logical tests in a single second. And there are 10^{45} processors in a cubic-meter computer. The computer should be able to handle 10^{68} tests per second.

That's fast. It is so fast that, in the first second, the computer could do all the necessary comparisons to build the list up to 225 beliefs.

And then, suddenly, things would slow down. It would take a second to add the 226th belief; two seconds to approve a 227th belief; about a minute to check the 232nd. The computer would be working as fast as ever, but the number of tests doubles as each belief is added to the list. It would take over a month to approve the 250th belief. Expanding the list to 300 would take—gulp!—*38 million years.*

Okay, but this is a thought experiment, and we have all the time in the world. The age of the universe is estimated to be about 10 billion years. That's between 10^{17} and 10^{18} seconds old. Add another order of magnitude or two (10^{19} seconds), and you have a decent approximation to "forever." By the time the universe is ten times older than it is now, virtually all the stars will have burned out, and life will probably be extinct. So 10^{19} seconds is about the longest amount of time that it makes any sense to talk about. It follows that if an ideal computer with 10^{45} processors worked from the beginning of time to its end, it could check the stupendous number of 10^{19} times 10^{68} subsets against new beliefs. That's 10^{87}. That's enough to take us up to a list of 289 beliefs.

We need a more powerful computer. Once the rock-bottom size of components has been reached, computers must get larger to get more powerful. Let the computer expand beyond the confines of a room, or a house . . . or a county or continent. However big it got, the ultimate limit would be the size of the universe.

The distance of the most remote quasar currently known is estimated at 12 to 14 billion light-years. If the universe is finite, a generous estimate of its "diameter" might be 100 billion light-years. A light-year is just under 10^{13} kilometers, or 10^{16} meters. That makes the diameter of the universe something like 10^{27} meters and its volume about 10^{81} cubic meters.

Therefore, a computer as big as the universe could contain 10^{45} times 10^{81} components the size of protons. That comes to 10^{126} components. Call this pipe dream absurd; the point is this: No matter what technical advances may await, no computer will ever be made of more than 10^{126} parts. No brain, no physical entity of any kind, could have any more parts. That is one limit we must live with. And if this computer works from the beginning of time to the end, it could execute at most 10^{126} times 10^{42} fundamental operations—10^{168} in all.

This 10^{168} is an absolute limit on how many times you could do *anything*. It's the closest thing to a supertask there is. There isn't enough time *or* enough space to allow more than 10^{168} of anything. And unfortunately, running 10^{168} logical tests still doesn't get us very far. The computer would conk out after extending the list to about 558 beliefs.

We can at most know 558 things?! No, of course not. We know many things through simple deductions, syllogisms, and sorites. The number 558 is the rough limit for beliefs logically complex enough to require an exponential-time-checking algorithm. A set of

558 beliefs as "unruly" as those in Carroll's pork-chop problem would probably exceed the computing power of even a computer as big as the universe. That is why new paradoxes continue to be invented.

Logically complex beliefs are not rare or unnatural. Even those beliefs we idealize as simple (like "All ravens are black") are actually qualified with a battery of auxiliary hypotheses. The difficulty of SATISFIABILITY speaks of more than logic puzzles.

When we cannot even tell if our more complex beliefs contain a contradiction, we don't fully understand them. We certainly can't deduce all that may follow from those beliefs. If you think of logical deduction as a kind of vision through which we see the world, then that vision is limited. Sorites, chains of simple deductions, are our principal lines of sight. Through them we peer far into the murk. Our vision for more complex deductions is extremely nearsighted. We don't see everything, not even everything implicit in our experience. There are things going on out there that we will never appreciate.

It is not even that we are too feebleminded to understand all that we miss. If we met up with the omniscient being who keeps popping up in these paradoxes, he would be able to show us what we are missing, and we could convince ourselves it was true. The answer to a puzzle is simple once you see it.

The "we" here includes humans, computers, extraterrestrial beings, and any physical agency. NP problems are hard for all. Stockmeyer and Meyer's thought experiment is an information-age counterpart to Olbers's paradox. From the fact that we see stars in the sky—from the fact that the entire universe is *not* a computer—we can be certain that no one in the universe knows everything.

PART THREE

10

MEANING

Twin Earth

THE VOYNICH MANUSCRIPT is a very old, 232-page illuminated book written entirely in a cipher that has never been decoded. Its author, subject matter, and meaning are unfathomed mysteries. No one even knows what language the text would be in *if* you deciphered it. Fanciful pictures of nude women, peculiar inventions, and nonexistent flora and fauna tantalize the would-be decipherer. Color sketches in the exacting style of a medieval herbal depict blossoms and spices that never sprang from earth and constellations found in no sky. Plans for weird, otherworldly plumbing show nymphets frolicking in sitz baths connected with branching elbow-macaroni pipes. The manuscript has the eerie quality of a perfectly sensible book from an alternate universe. Do the pictures illustrate topics in the text, or are they camouflage? No one knows.

A letter written in 1666 claims that Holy Roman Emperor Rudolf II of Bohemia (1552–1612) bought the manuscript for 600 gold ducats. He may have bought it from Dr. John Dee, a glib astrologer and mathematician who traveled on the winds of fortune from one royal court to another. Rudolf thought the manuscript was written by the English monk and philosopher Roger Bacon (c. 1220–92).

Bacon was as good a guess as any. As "Doctor Mirabilis" he had become a semi-mythic figure in the generations after his death, part scholar and part sorcerer. Bacon was a collector of arcane books. He knew about gunpowder and hinted in his writings that he knew about other things he wasn't ready to make public. At the time of his death, Bacon's works were considered so dangerous that (according to romantic conceit) they were nailed to the wall of Oxford's library to molder in the wind and rain.

The Voynich manuscript is said to have languished for a long time at the Jesuit College of Mondragone in Frascati, Italy. Then in 1912 it was purchased by Wilfred M. Voynich, a Polish-born scientist and bibliophile. Voynich was the son-in-law of George Boole, the logician, and husband of Ethel Lillian Voynich, one of the best-known English writers in the Soviet Union and China (for *The Gadfly*, a revolutionary novel long forgotten in the West). Lacking any intelligible title, the manuscript took on Voynich's name. Voynich brought it to America, where it was intensively studied. Scholars and crackpots have analyzed, then forgotten about, the Voynich manuscript in several cycles over the past seventy-five years. The manuscript is now at Yale University's Beinecke Rare Book and Manuscript Library.

The manuscript's cipher is no ordinary one. If it had been, it would have been cracked long ago. The cipher does not use Roman or any other conventional letters or symbols. It is not mirror-image writing or any simple distortion of familiar letters. The cipher employs approximately twenty-one curlicued symbols that loosely suggest some Middle Eastern scripts. Of course, the symbols aren't from any known Middle Eastern alphabet. Some symbols are joined together like slurred musical notes. A few symbols appear rarely—or maybe they are sloppy variants of the others. The writing forms "words" with spaces between them.

The diagram shows the commoner Voynich symbols labeled according to a scheme used by physicist William Ralph Bennett, Jr., who has subjected the manuscript to computer analysis. Bennett's letters (shown below each Voynich symbol) are arbitrary and serve only to name the symbols and allow computer entry.

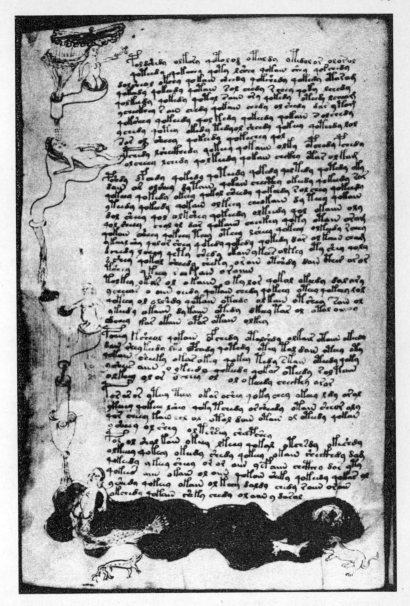

Folio 79 Verso from the Voynich manuscript

a)

ailmwno cċc ƤⲮⲪⲯⲮ ꙅꙅ᠕ⲅ ꙅ48ꙅ
A ILMNO   CET   PHFK   QUVY   ZDSG

b)

cc   ċc   cȼc   cⲮcc   cȼc   ccꙅcc
CT   ET   CPT   EHT   CFT   CHT ZCHT

a) The commonest Voynich manuscript symbols and the alphabetic labels assigned by William Ralph Bennett, Jr.
b) Common linked symbols with Bennett's transliterations

Some symbols (those labeled *A, I, L, M, N,* and *O)* are similar to lowercase script versions of the corresponding Roman letters. According to Bennett, other symbols look like letters of the Cyrillic, Glagolitic (old Bulgarian), and Ethiopian alphabets. The symbol labeled *Y* looks Chinese.

To add to the confusion, folio 17 contains a tiny note in Middle High German, not necessarily by the original author, talking about the Herbal of Matthiolaus. Some astrology charts in the manuscript have the months labeled in Spanish. What appears to be a cipher table on the first page has long since faded into illegibility.

About 40 pages of the manuscript are missing. Originally it contained 17 quires of 16 pages each. The last pages of the book contain pictures of stars in the margins, but no writing. That hints that the pictures were done first and the writing filled in later. In that case the pictures may be merely decorative. Many have nevertheless tried to read meaning into the illustrations. Some have guessed that the number of stars, women, or flowers on a page encodes something. The pictures of plants could mean the text discusses medicinal or magical uses of herbs, maybe an elixir of life or some such thing. Botanists have met with mixed success in identifying the plants depicted. The picture on folio 93 may or may not be a sunflower. A fruit on folio 101 suggests a capsicum pepper. Both are American plants unknown in Europe until Columbus returned in 1493. That was a couple hundred years after Roger Bacon's death.

All sorts of exotic scenarios have been suggested. The manuscript is in a lost dead language; it purposely avoids the most common letters of the source language to frustrate decipherers; it is a meaningless forgery (by John Dee? the Jesuits? Voynich?) created for monetary gain; it is the work of a medieval James Joyce, who invented his own words; it is the furious ravings of a long-forgotten madman. The Voynich manuscript is reminiscent of (and conceivably inspired) Borges's short story "Tlön, Uqbar, Orbis Tertius." In the story an eccentric millionaire financed a conspiracy of scholars to write an encyclopedia of the imaginary world "Tlön." The first drafts were in English, but the plan was to translate the encyclopedia into Tlön's (equally imaginary) language and calligraphy, creating a wholly inscrutable work.

The Voynich cipher has become a sword in the stone to cryptographers. Many of the most talented military code breakers of this century have tried to decipher it as a show of prowess. Herbert Yardley, the American code expert who solved the German cipher in World War I and who cracked a Japanese diplomatic cipher without knowing the Japanese language, failed with the Voynich manuscript. So did John Manly, who unscrambled the Waberski cipher, and William Friedman, who defeated the Japanese "purple code" of the 1940s. Computers have been drafted into the effort in recent years, to no avail.

The fact that computers have failed to decipher the Voynich manuscript may surprise some. As a practical matter, the task of cracking a cipher is mainly one of finding "weak spots." Just as a diamond is cut along dislocations in the crystal, ciphers are broken by exploiting telltale regularities. The Voynich manuscript appears to be an intractable cipher, a string of symbols purged of all the usual statistical earmarks of language. Attempts to solve it have been as useless as trying to drive a chisel through the geometrically perfect part of a diamond crystal.

Unless it is a fabrication, (and we will see later that it is almost certainly not), the text of the Voynich manuscript meant *something* to its author. It meant that something in part because of what the author was thinking at the time he or she wrote it. But is the meaning also inherent in the pattern of symbols? Or in a lost key to the cipher? Or in a combination of all of these? The possibility of our deciphering it depends on the meaning being "contained" in the pattern of symbols as well as in the now unknowable mental processes of the author.

It is doubtful that Bacon or any other medieval author singlehandedly created a cipher more secure than the scores of military codes of later centuries that *have* been cracked. Some see this as evidence that the Voynich manuscript is nonsense. A set of symbols doesn't *have* to mean anything. Is there any way of telling if a collection of symbols contains a message? This question is one of the most difficult ones in the study of knowledge.

Imagine someone in a far-future age digging up a time capsule containing a newspaper from our era. By then, English is a forgotten language; even the Latin alphabet is unknown. One archaeologist looks at the newspaper and decides it must be a form of writing. He hopes to decipher it and learn about the lives of the people who buried the time capsule. A second archaeologist says, "Don't waste your time! That's wallpaper! People glued it to the walls of their houses. Those little black squiggles are a decorative design that was popular back then."

You might think the first archaeologist would have an easy time demonstrating that the newsprint is writing and not a wallpaper design. There would be regularities in the newsprint—common letters, common words, periods at the ends of sentences—that would mark it as writing. The trouble is, there are regularities in decorative motifs too. It is hard to say offhand just how the regularities in an unknown design *necessarily* differ from those in an unknown script. The more alien the script or decorative art, the less confident one could be of any such determination.

Nor could the archaeologist necessarily expect to decipher the newspaper and prove his case that way. Egyptian hieroglyphic writing was never "deciphered" on the strength of internal clues. Only the serendipitous Rosetta stone revealed it to the modern world.

There is a bittersweet appeal to the puzzle of the Voynich manuscript. It's not just the prospect of discovering a medieval diary, a magic text, or a forbidden book of erotica. By its very inscrutability, the Voynich manuscript is a comment on the frailty of knowledge.

Roger Bacon

A pair of Bacons pioneered the scientific method: Franciscan monk Roger Bacon of the thirteenth century and Elizabethan statesman Sir Francis Bacon of three centuries later (1561–1626). Roger Bacon is by far the more mysterious of the two. Little is known of his life outside of what may be conjectured from his writings. We know he was an educator who lectured at Oxford and Paris. At some point in his career, he joined the Franciscan order and took a vow of poverty.

About 1247 Bacon grew dissatisfied with the faith his contemporaries had in Aristotelian science. He felt that direct observation
and experiment was superior to reliance on established authorities.
He credited this emphasis on experiment to Durand de Saint-Pour-
çain, a French Dominican philosopher of whom little else is known.
In 1267 Bacon reported having spent more than 2000 Parisian
pounds over the years on experiments and "secret books." From
one of these rare books he learned the formula for gunpowder. He
described the preparation of explosives in cryptic form.

Bacon's originality created friction between him and the Franciscan hierarchy. Fortunately, Bacon had befriended the man who
became Pope Clement IV. Clement, on hearing of Bacon's ideas for
a philosophical encyclopedia, ordered Bacon to send him a copy.
The Pope thought that the work already existed. Actually, it was
just an idea that Bacon had partially sketched in letters to friends.
Rather than explain, Bacon set to work. He concealed the project
from his Franciscan brothers, working without copyists. A year and
a half later he had a trilogy: *Opus Majus, Opus Minus,* and *Opus
Tertium.*

These works made Bacon notorious for speculations on future
technology. He described a telescope (but did not have a practical
model). He envisioned automobiles and, less accurately, airplanes.
Bacon was thinking of human-powered flight, in which human arms
flapped artificial wings. He also concluded that balloons might be
made to float by filling them with gases lighter than air.

Bacon believed the earth was round. *Opus Majus* described a sea
voyage west from Spain to India. Cardinal Pierre d'Ailly plagia-
rized this passage in his *Imago Mundi* (published 1480), where Columbus read it and cited it in a letter to Ferdinand and Isabella of
Spain.

Eventually Bacon's reputation as a miracle worker overwhelmed
him. The Franciscans imprisoned him around 1278 for "suspected
novelties." The story about Bacon's enemies destroying his books
after his death is apparently false. As far as we know, all his major
works survive.

False Decodings

The Voynich manuscript has driven some if not to madness then
to extraordinary self-delusion. More than one person has gone to
his grave *thinking* he decoded the Voynich manuscript.

In 1921 Professor William Romaine Newbold of the University of

Pennsylvania announced that he had deciphered the Voynich manuscript and would reveal his findings at a meeting of the American Philosophical Society. Like many, Newbold attributed the manuscript to Roger Bacon. He thought it proved that Bacon had built both a microscope and a telescope centuries before Galileo and van Leeuwenhoek. The illustration on folio 68, Newbold thought, was the spiral nebula in Andromeda as seen through Bacon's secret telescope. Newbold even reported that the telescope's mirrors had cost Bacon the equivalent of $1500. Other illustrations showed spermatozoa and ova. Newbold's revelations briefly excited the press and public. One woman was so sure that Newbold had uncovered Bacon's black incantations that she traveled hundreds of miles to ask Newbold to cast out the demons that had possessed her.

It is now sadly clear that Newbold himself was possessed. He was reluctant at first to give away too much of his discoveries. The more he made public, the more it became obvious that he was reading his own hunches into a still uncracked cipher. Newbold had Bacon observing the spiral structure of the Andromeda nebula with a reflecting telescope. Astronomers pointed out that the nebula's spiral structure isn't visible in *any* telescope—only in time-exposure photographs. Even Newbold didn't have Bacon inventing the camera. From the earth, the Andromeda nebula is seen almost edge on. Whatever is being depicted in folio 68 is face on, so that its outline is a circle.

The cipher Newbold attributed to the manuscript was a masterwork of wishful thinking. He found a barely legible "key" on the last page of the manuscript. (More than one scholar has supposed that this inscription may be a key. Some think the handwriting of the "key" is different and that this inscription was added later by someone other than the author.) Newbold claimed that the symbols translated to the Latin *A mihi dabas multos portas* ("Thou wast giving me many gates"). This he took to mean that more than one cipher was used.

According to Newbold, Roger Bacon encoded an original Latin text with a "biliteral" cipher. In a biliteral cipher, a pair of letters in the visible writing encodes one letter of the message. This was ingenious by the standards of thirteenth-century cryptography, and should have ensured the secrecy of whatever the author was writing.

But that, insisted Newbold, was just one of a series of cryptographic Chinese boxes. In a normal biliteral cipher, the encoded message ("ciphertext") is twice as long as the original message ("plaintext"). To make the ciphertext more concise, Newbold thought that Bacon chose letter pairs so that the last letter of one pair was always the same as the first letter of the next pair. If Bacon wanted to encode the Latin word *unius,* he might have *or* representing *u,* and *ri* meaning *n,* and so on:

U   N   I   U   S
OR RI IT TU UR

Then he would strike out the repeated letters and get *oritur.* To make the cipher yet more baffling, more than one letter pair could encode a given letter, and letters that sounded similar—such as *b, f, p,* or *ph*—could be encoded by the same letter pair.

Confused? So was Newbold's audience. Those cryptographers who were still following what Newbold was saying realized that such a cipher would be hopelessly impractical.

There was more. If any letter pair contained one of the letters in the word *conmuta,* it was subjected to a further coding process that Newbold called "commutation" but never fully explained. Next the whole message was scrambled so that it was an anagram (!) of the prior stage.

Then came the topper: The visible symbols of the manuscript were just a cover, Newbold said. They didn't mean anything. Newbold believed that if you examined the symbols with a magnifying glass, you would find them to be made out of about ten tiny separate strokes each. He supposed that Bacon had used his newly invented microscope to create these tiny symbols. These tiny strokes were symbols of ancient Greek shorthand. The real message was in this microdot Greek shorthand. To decode the manuscript, then, you had to transliterate these microscopic symbols into letters, and *then* reverse the Byzantine process of anagramming, commutation, and assigning letter pairs.

Newbold's microscopic symbols were as fleeting as the canals of Mars—more so, for Newbold was the only one who saw them. To the extent that they had any reality outside Newbold's head, they were the irregularities of a coarse ink on rough paper.

Had the manuscript's author used Newbold's method, it would have been maddening indeed to encode anything. And once encoded, there would be no reliable way of decoding it. What you got could always be an *anagram* of letters that *sound like* the real message, etc., etc., etc.

Newbold's comments on his microscopic symbols are a poignant monument to self-deception. He wrote:

But the difficulty of reading the cipher characters is very great indeed. When first the letters were written they were, I think, distinctly visible under the proper degree of magnification, but after the lapse of more than six hundred years the writing on many pages has been so injured by fading, scaling, and abrasion, that the characters can scarcely be seen at all. In the second place, much depends upon the degree of magnification which was used by Bacon at the time of writing. The line which to the naked eye seems quite simple, when magnified three or four or five diameters is frequently seen to be composed of individual elements, and if it be magnified still further some of the elements will be resolved into still other elements, many of which may be taken as characters. . . . Another very great difficulty is that offered by the elusiveness of the characters themselves. The differences between them are very slight; when they are written under a microscope, even Bacon's own hand often gives to the differences but faint and ambiguous expression. Furthermore the characters are so interwoven one with another that it is often all but impossible to disentangle them. . . . I frequently, for example, find it impossible to read the same text twice in exactly the same way.

Stranger yet is American physician Leo Levitov's recent alleged solution of the Voynich manuscript. In 1987 Levitov claimed that the manuscript is in an unknown European language used by a cult of Isis worshipers around the twelfth century. Levitov believed that all other traces of the cult had been destroyed by the Spanish Inquisition. Levitov offers the most gruesome exegesis of the illustrations yet. The cult believed in euthanasia by opening a vein in a warm bathtub, and the pictures of enigmatic bathers show devices for draining off blood!

Levitov's peculiar language consists of twenty-four verbs and four pronouns of mutable spelling. His chaotic and uniformly morbid translations ("ones treat the dying each the man lying deathly ill the one person who aches Isis each that dies treats the person," begins folio 1) do not inspire confidence.

There is a certain pathos in these "decipherments." We all interpret language and even experience in a way that is both complex and difficult to describe. It is not that Newbold and Levitov are undeniably wrong. It is at least *conceivable* that the supposed authors wrote just what they thought and encoded it just as they said.

Most rational people do not ponder Newbold's and Levitov's cases long before rejecting them. To say exactly *why* we reject them is something else. Susan Sontag defined intelligence as a "taste in ideas." It is difficult to codify that taste.

Sense and Gibberish

The relationship between cryptographic problems and the experimental method has often been remarked upon. Cryptographer John Chadwick wrote:

> Cryptography is a science of deduction and controlled experiment; hypotheses are formed, tested and often discarded. But the residue which passes the test grows until finally there comes a point when the experimenter feels solid ground beneath his feet: his hypotheses cohere, and fragments of sense emerge from their camouflage. The code "breaks." Perhaps this is best defined as the point when the likely leads appear faster than they can be followed up. It is like the initiation of a chain-reaction in atomic physics; once the critical threshold is passed, the reaction propagates itself.

For the sake of argument, suppose that the Voynich manuscript was written by a clever con man and is completely without meaning. There *does* seem to be a simple way of telling whether it's gibberish, even without deciphering it.

The work of cryptographers depends on the statistics of language. Not all letters are equally common. With many types of ciphers, this means that the visible symbols have different frequencies.

The commonest letter in English is *e*. It is not everywhere the commonest letter (it's *o* in Russian), but every natural language favors some letters over others.

You might think a forger picking meaningless symbols at random would fail to favor some over others. Not necessarily. Try writing a "random" string of letters or numbers. It is very difficult not to favor certain letters or numbers unconsciously. True randomness is all but impossible for the human mind to create. A forger might happen to favor some symbols in a way that would approximate the letter frequencies of his native language or some other language.

That does not mean that the statistical approach is useless. There are more subtle considerations. In a real letter-substitution cipher, certain pairs of letters ought to be more common than others. For instance, *th* and *is* are quite common in English, and *q* is almost certainly followed by a *u*.

It works the other way too. Some pairs of letters are relatively uncommon. The letters *c* and *d* are common, but you rarely see *cd* in English text. The same principles apply to triplets of letters or larger groups. All the vowels are common, and many pairs of vowels are common, but most instances of three consecutive vowels are rare or nonexistent.

That this does provide a means of distinguishing a real cipher from nonsense is demonstrated by the false cipher in Balzac's *The Physiology of Marriage*. Published in 1829, *The Physiology of Marriage* is a satirical handbook on marriage and adultery. Inserted after the words "L'auteur pense que la Bruyère s'est trompé. En effet, . . ." is a two-page cipher that has never been deciphered. Many readers tried to decipher it, goaded by the suspicion that the passage must have been so scandalous that the publisher dared not print it as written. Balzac dropped hints about it for years after the book was published.

The cipher contains uppercase and lowercase letters, many with accents and some upside down. There are numerals and punctuation marks, but only a few blank spaces. It was significant, some thought, that the cipher ends with "end" and contains the exclamation "sin!" (both in English).

The statistics of the Balzac cipher are wildly at variance with French or any other European language. In this case there is little doubt that the symbols were picked at random, probably by the typesetter. Some later editions of the book even have different "ciphers."

The Voynich manuscript has been subjected to similar scrutiny. Unlike the Balzac pseudo-cipher, the Voynich symbols have statistical patterns much like real languages. There are pairs of symbols that often go together (*AM, AN, QA,* and *QC,* according to Bennett's labels). There are common symbols that are rarely combined. In fact, these patterns are even more pronounced than in English. The Voynich text is less "random" than any known European language.

A statistic called "entropy" measures the degree to which letters or other symbols form recurring patterns in the text. By a weird coincidence, the entropy per symbol of the Voynich manuscript is about that of Polynesian languages. None of the many suppositions about the manuscript had it enciphered from Hawaiian or Tahitian.

The Polynesian languages are known for their economy of letters. The Hawaiian alphabet has just twelve, plus a heavily used apostrophe. The Voynich manuscript uses twenty-one common symbols plus a few rare ones. The manuscript's entropy suggests that its source text was much more ordered than most natural languages.

That is strong evidence that the Voynich manuscript is a real cipher, not gibberish. It is hard to believe that a forger was sophisticated enough to simulate the statistics of language.

It also confirms that the text is not a simple encipherment of any European tongue. The manuscript seems to be in a "language" with fewer common letters than European languages. Perhaps the author lumped similar-sounding letters together, somewhat as Newbold conjectured. Or the plaintext could be an Esperanto-like language of the author's invention. The bulk of contemporary scholarly opinion thinks the manuscript was written after Columbus's return (not by Bacon, obviously).

The Parable of the Cave

Many of the questions raised in cryptography are far from mundane. The way we interpret experience is much like the unraveling of a cipher. Is our mental picture of the world inherent in the stream of sensory experience, or is it largely in a key, in a way our brains translate this experience?

A classic forebear of the thought experiments in this book is the Parable of the Cave in Plato's *Republic*. Book VII begins with this dialogue between Socrates (the first speaker) and Glaucon:

> And now, I said, let me show in a figure how far our nature is enlightened or unenlightened: — Behold! human beings living in an underground den, which has a mouth open towards the light and reaching all along the den; here they have been from their childhood, and have their legs and necks chained so that they cannot move, and can only see before them, being prevented by the chains from turning round their heads. Above and behind them a fire is blazing at a distance, and between the fire and the prisoners there is a raised way; and you will see, if you look, a low wall built along the way, like the screen which marionette players have in front of them, over which they show the puppets.
>
> I see.
>
> And do you see, I said, men passing along the wall carrying all sorts of vessels, and statues and figures of animals made of wood and stone and various materials, which appear over the wall? Some of them are talking, others silent.
>
> You have shown me a strange image, and they are strange prisoners.
>
> Like ourselves, I replied; and they see only their own shadows, or the shadows of one another, which the fire throws on the opposite wall of the cave?
>
> True, he said; how could they see anything but the shadows if they were never allowed to move their heads?

And of the objects which are being carried in like manner they would only see the shadows?

Yes, he said.

And if they were able to converse with one another, would they not suppose they were naming what was actually before them?

Very true.

And suppose further that the prison had an echo which came from the other side, would they not be sure to fancy when one of the passers-by spoke that the voice which they heard came from the passing shadow?

No question, he replied.

To them, I said, the truth would be literally nothing but the shadows of the images.

The relationship between our mental images of the world and the external reality continues to fascinate and vex. Several modern paradoxes twist this correspondence to the breaking point.

The Electronic Cave

Many technological versions of Plato's scenario are possible. Imagine a cave-bound prisoner who watches the outside world on a closed-circuit TV screen. As in Plato's allegory, this prisoner has been chained to the cave wall from birth. A video camera outside the cave constantly transmits pictures to the prisoner's TV. Furthermore, the prisoner's head is in a swiveling harness. As he moves his head to the right, the TV swings to the right on silent, perfectly balanced bearings so that the screen always fills the visual field. Outside, the camera rotates a like angle so that the field of view on the screen changes in what must seem a perfectly natural fashion to the prisoner.

With this setup, the cave dweller's captors could play even stranger tricks on his perception. What if, unknown to the prisoner, the TV camera forever shoots into a mirror at a 45-degree angle? Everything the prisoner saw would be reversed right to left. The cave dweller would be unaware that he was seeing a mirror image of reality. If he learned to read books placed in front of the camera, he would learn to read backwards.

The TV image could be permanently upside down. Again, the cave dweller would think the way he saw the world was the right way. The fact that his TV image was upside down would be no more relevant than the fact that the images on our retinas are physically upside down. As long as the prisoner spent his whole life looking at an upside down image (a right-side-up retinal image) he could no more be aware of the reversal than a fish is aware of water.[1]

The cave dweller's severest limitation (both in these variants and in Plato's original story) would be the lack of feedback. The cave dweller could not push something and watch it move in response. He could not observe any of the thousands of other ways that one's will may modify the environment.

The cave dweller might be given a more active role with robotic technology. His arms and legs could be fitted with sensors. The movements of the real limbs would be relayed to a robot body outside, situated near the TV camera. As the cave dweller lifts his finger, the robot finger moves, much as in a radioisotope lab. With sophisticated robotics, this could mimic the brains-in-vats situation. The cave dweller would think he inhabited the robot body in the outside world. He would have no way of knowing he was actually in a cave, and would be skeptical of any such claim.

One question raised by these fantasies is how much "information" is necessary to create a mental image of a world. Here's a more extreme predicament. Let the prisoner's TV screen be a video-text terminal. Rather than pictures, a running commentary (in written English) of what is happening outside the cave fills the screen at all times. Never does the prisoner see a bird or even a TV image of a bird. Instead a message like "A BLUEBIRD JUST PERCHED IN THE TREE OUTSIDE THE CAVE OPENING" scrolls across his screen.

Now the situation resembles that of the decipherers of a lost language. Again, the cave dweller has no experience of the outside world at all. He has been strapped to the wall of the cave since birth. He cannot know how to read *unless* he can puzzle out the meaning of the strange symbols filling his visual field. Is that possible?

The prisoner would probably learn a lot about the structure of written English. He would know the shapes of every letter and punctuation mark by heart. He would recognize short common words. Nothing else demands his time in the cave; much of his imagination would be directed to the screen hieroglyphs. Maybe he

[1] In 1928 Theodore Erismann of the University of Innsbruck had human volunteers have their vision weirdly distorted with special glasses. Over several weeks of constant wear, subjects adjusted to glasses that reversed the visual field top to bottom or left to right and to a mirror device that allowed the wearer to see only what was behind his head. One subject rode his motorcycle through city streets wearing right-to-left goggles. All had to readjust when they stopped wearing the goggles.

would come to know all the common words without making an effort, just as a rancher may recognize, though not know the names for, dozens of kinds of wildflowers.

Learning the meaning of words is something else again. Language is based on shared experiences. There is no way of describing the color red to a person blind from birth. The poverty of the cave dweller's life would allow few points of reference.

Conceivably, the prisoner might notice that the sentence "THE SUN IS RISING" appears at regular intervals. With enough patience, he might notice that some words or phrases (like "MORNING," "OWL," "FULL MOON," and "SNOW") appear on the screen at certain times and not at others. These temporal clues might provide a starting point for deducing the meaning of the writing (though it seems unlikely the prisoner would get very far).

The Binary Cave

The absolute minimum amount of information that can be conveyed about anything is a simple yes or no—the "bit" or binary digit of computer science. Consider this ultimate version of the parable: Outside the cave is a TV camera as before. It converts images of the landscape into electrical impulses. Information about brightness, hue, saturation—all the pellucid effects of a rainbow—are coded as a sequence of 1's and 0's that have meaning to digital video equipment: 010110100110010101101100110111 . . . This information travels by cable into the cave. In this case there is no TV screen in the cave. Instead the stream of incoming bits is fed into a much simpler display device. When the device receives a 1 in the input, it projects a dot of white light on the wall of the cave in front of the prisoner. When it receives a 0, the wall is dark. The result is a flickering dot of light on the cave wall. The prisoner spends his entire life watching that flickering dot and nothing else.

In a strong sense, we are *all* like this prisoner of the cave. Our entire experience of the world is a sequence of nerve impulses that could be expressed as a sequence of 1's and 0's. What is amazing is that we can draw any conclusions whatsoever from such an impoverished input. Everything, from the fact that space has three dimensions to World Series predictions, is derived from that same abstract input. That is the puzzle of knowledge: how we derive meaning from symbols that seem incapable of expressing anything of the variety of the world.

Next to this denizen of the cave is another. His situation is apparently the same. His entire life is spent watching a flickering dot encoding a sequence of 1's and 0's. By the same miracle of imagination, he has built a rich mental picture of the outside world: its geometry, its eons of past history, its distant future. But because of an equipment malfunction that has caused his dot of light to go on and off at random, this picture is *completely wrong*.

Can a Brain in a Vat Know It?

Now presumably this is impossible. To put it in a slightly different context, suppose there are two brains in vats. Brain A receives a carefully modulated stream of impulses to create the illusion of a world, and brain B receives a random stream of impulses because of a hardware error. Surely there is something "worldlike" in A's input, and not in B's random input. Brain B would not be able to make sense of its input at all. The meaning is inherent in A's input —so we would guess.

Several philosophers have used brains-in-vats thought experiments to explore the issue of meaning. In *Reason, Truth and History* (1981), Hilary Putnam controversially argues that we are *not* brains in vats and can know it. This brought howls of protest from the philosophic community. His reasoning is clever and does not quite assert what it seems.

Assume, as the premise of a reductio ad absurdum, that we are brains in vats. Then when we say "bowling ball" (and of course we don't really *say* anything, having no lips) we refer not to physical round objects with three holes in them—of which there may be none. (There may be no bowling alleys in the "real" world outside the brains-in-vats laboratory.) Even so, "bowling ball" refers to *something*. It refers to a certain pattern of electrical stimulation by which the laboratory's mad scientists create the illusion of bowling balls. This is the physical counterpart of the thought, the object of the reference.

You might say there are two languages, vat-speak and lab-speak. "Bowling ball" in lab-speak refers to the round thing with three holes. "Bowling ball" in vat-speak refers to an electrical impulse that creates the image of a round thing with three holes.

If "bowling ball" refers, in vat-speak, to electrical impulses, what does "brain" refer to? Not to a gray lump of neurons, but to another set of electrical impulses, the impulses by which the illusion of physical brains is created. "Vat" refers to an electrical stimulation too. So if we are brains in vats, the words "brains in vats" refer not to physical brains in physical vats but to one type of electrical stimulation "in" another type of electrical stimulation. To say "Yup, I'm a brain in a vat" is wrong because we are not electrical impulses, we are *real* brains in *real* vats!

Putnam therefore sees the statement "I am a brain in a vat" as necessarily false. It is perhaps comparable to the statement "The universe rests on the back of a big turtle." You can maintain that this is necessarily false too, because "universe" means everything, including the turtle if it exists. The universe can't rest *on* anything else because there *is* nothing else—by definition.

Defensible as this is, it neglects the flexibility of language. The average person would understand "The universe rests on the back of a big turtle" to mean that the known universe of stars and galaxies rests on an unknown turtle. We automatically redefine "universe" to fit the context of the sentence—as we would with a statement like "We are brains in vats."

A brain in a vat *might* be able to express the "real" state of affairs and still satisfy semantic purists. It would have to recognize the difference between vat-speak and lab-speak and assert something like "I am what 'a brain in a vat' means in lab-speak." Arguably (?) this avoids the same problem of misplaced reference because "lab-speak" is a metaphysical term of vat-speak with no physical correlative.

Twin Earth

Putnam's best-known thought experiment challenges the idea that meanings are "in the head," a matter of mental states. Suppose, Putnam says, there is another planet in our galaxy called Twin Earth. Twin Earth is exactly like Earth in almost every way. There are regular-looking people on Twin Earth, and they even speak English (as in a lot of science-fiction movies). Twin Earth is so incredibly similar to Earth that they call their planet "Earth." (It would be silly of them to call it "Twin Earth"!)

One difference between Earth and Twin Earth is this: Oceans, rivers, lakes, raindrops, and tears on Twin Earth are made out of a transparent liquid that looks exactly like water but isn't water. That is to say, it isn't *chemically* water. Instead of H_2O, it has a different formula we'll write as XYZ. But so amazing is the "parallel evolution" of Twin Earth that they still call this liquid "water." When an inhabitant of Twin Earth talks of watering the lawn, he means putting XYZ on the lawn. Putting H_2O on the lawn might kill it.

There is H_2O on Twin Earth, a few tightly stoppered bottles of it in chemistry labs, but they call it something other than "water." There is XYZ on Earth. Chemists on both planets can distinguish the two compounds with a simple test.

Now think what would happen if, a few centuries from now, we sent a spaceship to Twin Earth. Our astronauts would get out of the ship, take off their space helmets, and introduce themselves in English to the natives. After a while, one astronaut would get thirsty and ask for some water. A Twin Earth host would go to the faucet and draw out a nice, tall glass of "water." The astronaut would put the glass to her lips, take a sip, and spit it out! The Earthlings would run chemical tests and find that the "water" on Twin Earth is toxic, undrinkable XYZ.

Now turn back the calendar to the year 1750—the year called "1750 A.D." on both Earth and Twin Earth, for they have identical calendars. There is no such thing as space travel. Our astronomers' crude telescopes have not yet picked out the star that Twin Earth orbits, and vice versa. Chemistry is in its infancy as well. Our chemists have not yet discovered that water is made of hydrogen and oxygen atoms. Twin Earth's chemists have not yet discovered that their "water" is made of X, Y, and Z.

In the Earth of 1750 there is an individual called Oscar. There is an extremely similar individual, also called Oscar, on Twin Earth. The two Oscars are so alike that they even have the same thoughts at all moments of their lives. When Earth's Oscar uses the word "water," it has precisely the same memories and mental associations as it does when Twin Earth's Oscar uses the word. Both think back to a certain water fountain on a school playground; to the first time they saw the Atlantic Ocean (a geographic feature common to both planets); to the water that leaks through the roof of their house in a heavy rain. If you were to ask the Earth Oscar to explain what water is, he would say it's such and such, and if you were to ask the Twin Earth Oscar, he would say exactly the same thing. There is nothing in the one Oscar's consciousness that distinguishes his understanding of water from the other Oscar's. Yet the two waters are not at all the same. Putnam concludes: "Cut the pie any way you like, 'meanings' just ain't in the *head!*"

But if meanings aren't in the head, where are they?

Twin Earth Chemistry

Many philosophers *do* believe that meaning is largely "in the head." A word like "water" could mean anything at all. It means what it happens to mean, what we are thinking of when we say "water," and not something that is dictated by the shapes of the letters. If we found a very short scroll in a lost language and all it said was "water," there would be no hope of translating it.

There was a time for all of us when we were unsure what "water" meant. Parents and other adults said "water" in certain contexts, and it was up to each of us to decide what was common to each context. As adults we think our experience is broad enough to eliminate that ambiguity. At this late date, could "water" possibly be anything other than what we think it is?

Two objections are often raised to Putnam's thought experiment. It is true, though not necessarily germane, that Twin Earth is biochemically implausible. Lest that concern get in the way, a word about the putative chemistry of Twin Earth may be in order.

In Putnam's 1975 article, he describes the hypothetical XYZ as a liquid "whose chemical formula is very long and complicated." It is said to be liquid at the same range of temperatures and pressures as H_2O, and of course it quenches Twin Earth thirsts and otherwise fulfills the ecological and biochemical role that our water does.

There is no known substance that is so nearly similar to water, without *being* water, as that. It is doubtful that any substance with a complex formula would fit the bill. Water's predominant role in Earthly biochemistry depends on the small size of its molecules. Those liquids with long, complicated formulas are generally oily, viscous, and otherwise unwaterlike.

Hydrogen peroxide (H_2O_2), the only other compound of hydrogen and oxygen, is wildly unstable and could not exist as oceans. (The "hydrogen peroxide" sold in drugstores is a very weak water solution of this compound.) Hydrogen sulfide (H_2S) is chemically analogous to water but is a gas. Looser analogues are gaseous ammonia (NH_3) and hydrogen fluoride (HF), a virulent acid whose boiling point is just below (Earthly) room temperature.

Science-fiction writers sometimes speculate about life on planets where ammonia takes the place of water in the biochemistry. Such planets would have to be much colder than Earth, for ammonia is a liquid only below temperatures of 36 degrees below o Fahrenheit

(−33 degrees C.). (The liquid "ammonia" sold for cleaning windows is again a water solution of gaseous ammonia.)

There are probably many Earth-sized planets or moons with the right range of temperatures for ammonia to be liquid. Unlike the other compounds mentioned above, ammonia is common (Jupiter has ammonia clouds) and might well form lakes, oceans, and rivers. Ammonia is a polar compound like water, meaning that it can dissolve a wide range of substances. This quality seems essential for any conceivable biochemistry.

That said, ammonia could not be Putnam's XYZ—not if Twin Earth is really all that similar to Earth. Look at a few minor points: What would Twin Earthers clean their windows with? They couldn't use just NH_3, because that's their "water," and if Twin Earth is so similar to Earth it ought to have a product with the trade name Windex that contains something other than "water." At the temperatures at which ammonia is liquid, mercury is solid. There could be no mercury thermometers or barometers; no dental fillings compounded with mercury. They couldn't call some other liquid metal "mercury" because *all* metals are solid at that temperature. Of course, this is the least of it. You don't have to follow any such train of thought long to see that there would be differences by the thousands. A biochemistry based on ammonia would probably preclude anything similar to the human race evolving (even if we admit the possibility of intelligent life based on ammonia).

Others seize on Putnam's figure of speech ("meanings ain't in the head"). The human body is mostly water. Not only do we say and think "water," but there is water in our head while we do it. If Twin Earth indeed has an XYZ-based chemistry, there would be XYZ "water" in the head of every Twin Earther. The meaning *is* in the head after all!

Although some disagree, I do not think either of the above objections substantially hurts Putnam's argument. Putnam's article gave some less dramatic examples that illustrate the issue just as well. He asked, what if the "aluminum" pans on Twin Earth were actually made of molybdenum, and "molybdenum" was really aluminum?

To a chemist, this example is not as compelling as it might be, for molybdenum is much, much heavier than aluminum and different in other important ways too. There are, however, elements with very similar chemical and physical properties. Many of the rare earth (rare twin earth?) elements are indistinguishable except by fairly sophisticated chemical assay. Moreover, they play no role in human nutrition. You could imagine, if necessary, a human or Twin Earther brain that contains not a speck of either element.

None of the rare earth elements are common enough to be familiar to nonchemists. A better-known case of near-twin elements is nickel and cobalt. Nickel and cobalt are indistinguishable in appearance and have almost exactly the same density and melting point. Both are among the few metals that can be magnetized, and they have similar chemical properties.

Suppose, then, that what they call "nickel" on Twin Earth is cobalt, and vice versa. The Twin Earth countries called "Canada" and "the United States of America" issue coins called "nickels." They're called that because they contain the metal "nickel," which is to say, cobalt. Twin Earth nickels nonetheless look identical to our nickels. Neither nickel nor cobalt plays much of a role in human nutrition, so astronauts on Twin Earth would not develop a deficiency of one or the other. It might well take some time for anyone to notice the difference.

Eventually, an astronaut with training in chemistry might look at a periodic table chart on Twin Earth and note that the symbols "Ni" and "Co" were seemingly swapped (though I imagine that someone who got good grades in college chemistry classes could easily fail to notice the difference too). Another tip-off: In everyday speech on Earth, the word "cobalt" more often means a shade of blue rather than the element. Cobalt blue, an artist's pigment, is made from cobalt oxide. There would be no cobalt blue on Twin Earth. That particular vivid, slightly greenish blue pigment would have to be called "nickel blue."

Putnam's thought experiment demonstrates that all experience is ambiguous. Both Oscars had identical experiences with water. The XYZ water may even taste the same to the Twin Earth Oscar as H_2O water does to the Earth Oscar. The very sequence of firings of the neurons in both Oscars' brains could be identical, yet there is more than one external reality compatible with them.

The Libraries of Atlantis

Let's suppose that another minor difference between Earth and Twin Earth is that Twin Earth has an extra continent called Atlantis. Atlantis has its own language, which has no linguistic affinities with the other Twin Earth languages (which, aside from a few problem words like "water" and "molybdenum," are identical with Earth languages).

An astronaut from Earth visits a library in Atlantis with an interpreter who speaks both English and Atlantean. The astronaut is surprised to find on the shelves a copy of *Gulliver's Travels* by Jonathan Swift. At least, that's what it seems to be. The cover bears those words, in English, in the Roman alphabet. Flipping through the book, the astronaut sees that it tells the familiar Swift satire in English. Another case of parallel evolution!

The astronaut comments to the interpreter that we have the same book by the same author on Earth. "Really?" says the interpreter. "It's based on a true story, you know."

"Now don't tell me there's really a place called Lilliput on Twin Earth!"

"What? Oh! No—the book you're holding is an Atlantean copy of a play, *Henry VI,* by an author called William Shakespeare. This confuses people all the time. When *Henry VI* is translated into the Atlantean language, it looks, *superficially,* like *Gulliver's Travels* in English."

It further turns out that the real *Gulliver's Travels* in Atlantean looks just like *The Grapes of Wrath* in English, and *The Grapes of Wrath* in Atlantean looks like the 1982 Tallahassee phone book. According to the interpreter, an English speaker and an Atlantean speaker can read the same book, and for one it will be *1001 Jokes for Toastmasters* and for the other a commentary on the Koran. Hence the Twin Earth saying: "Meanings just ain't in the book."

Is the interpreter pulling the astronaut's leg?

It is, of course, extremely unlikely that two languages would happen to bear the relationship described. The question is whether it is at all possible. All the books mentioned, except possibly the phone book, repeat many common words like "the," "of," and "a." It could be, for example, that "the" in English translates into "of" in Atlantean. Maybe every word in Atlantean is spelled the same as a (different) English word. Then a translation from English into Atlantean would indeed produce a jumble of "English" words. But they certainly wouldn't form meaningful sentences, and in any case, the pattern of repeated words in *Gulliver's Travels* is not the same as in *Henry VI*. A translation of *Henry VI* from English to any language could not produce *Gulliver's Travels*.

Not if the translation is word for word, anyway. But it is far from clear that the apparent words in Atlantean text *are* words. It could be that the space between "words" is really a letter in the Atlantean alphabet. And it could be that some "letter" is actually a null character inserted to separate words.

More to the point, most translation is not word for word. In some cases (as in English to German) that is impossible because of a different word order in sentences. There may be languages so alien to English that is necessary to translate a paragraph or more of the text at a time. It is then conceivable (if fantastically unlikely) that *any* book could be any other book in some alien language.

Cryptographic systems have yet more freedom than languages. To take an extreme example, it is possible that the Voynich manuscript encodes the text of the Gettysburg Address. How? One way is the following cryptographic system: "If you want to encode the text of the Gettysburg Address, make this sequence of squiggles: (insert Voynich manuscript here). If you want to encode anything else, just write it in Pig Latin." You can't *prove* this wasn't the enciphering system used to create the Voynich manuscript.

Poe's iiiii . . . Cipher

Edgar Allan Poe, an amateur cryptographer, once ran a magazine competition in which readers were invited to send in ciphers to be deciphered. In a follow-up article on cryptography, he mentioned the possibility of finding this string of letters in a cipher: *iiiiiiiiii*. How would you decipher that, knowing that no word of English has that many repeated letters?

A string of ten *i*'s is possible; it just couldn't arise in a simple cipher where one letter stands for a different letter throughout. You could even have a cipher in which the entire message was an unbroken string of the same letter.

For one thing, it could be a totally ambiguous cipher in which the letter *i* stands for all 26 letters and the encipherer relies on his ability to reconstruct the plaintext later. Or it could be a cipher in which a different letter-substitution rule is used for each letter of the plaintext.

In such a cipher all meaning has been purged from the ciphertext itself. You can demonstrate this by imagining that someone finds a cipher manuscript filled with *i*'s and nothing else. The finder claims that it can be deciphered into the text of the Gettysburg Address. This claim would be ridiculous. You could just as well claim that it meant anything else. The "meaning," if any, evidently resides in the cipher system or in the head of whoever wrote the plaintext. Since there is no meaning in the ciphertext, it is an unbreakable cipher.

Most familiar "codes" are really ciphers. The Morse "code" is a cipher; so are most "codes" of military importance. In a true code, symbols stand for ideas. The no-smoking glyph of a cigarette in a red circle with a slash through it is an example; so are the dozens of other international symbols seen in airports and other public places. A code attaches meaning to individual symbols.

It is difficult to express oneself in code. A code has symbols only for the common words and messages that have been anticipated by the designer of the code. Codes are awkward, often useless, in communicating the unexpected. For that reason the important military, espionage, and diplomatic "codes" are ciphers.

In a cipher, symbols stand for letters. A cipher lets you compose your message in plain English (or plain anything) and then convert it into symbols. The symbols are decoded by the recipient to recover the precise original message.

A cipher replaces one letter or other typographic symbol with another. Some rules are simple; some are more complex. Any letter-for-letter substitution can be represented by writing the alphabet (which here will mean all permissible symbols, including punctuation and numerals, if used in the cipher) in conventional order, and below it, the letters that are substituted. One simple substitution scheme is:

A B C D E F G H I J K L M N O P Q R S T U V W X Y Z

B C D E F G H I J K L M N O P Q R S T U V W X Y Z A

A becomes *B*, *B* becomes *C*, *C* becomes *D*, and so on. All the letters are encoded by their successors in a circular alphabet. The word *MESSAGE* becomes *NFTTBHF*. The process is easily reversed by the message's recipient.

Ciphers using this type of substitution throughout are called Caesarian, after their use by the Roman emperors. Augustus Caesar used the cipher above; Julius Caesar used the similar one in which plaintext *A* is replaced by *D; B* by *E*, etc.

There are 26 Caesarian ciphers. The bottom row of letters could have been shifted two letters, three letters, etc. (The 26th "cipher" merely replaces each letter with itself.) Each Caesarian cipher can be easily designated by a number or letter. You might designate the cipher by the ciphertext equivalent of the plaintext letter *A*. Then Augustus Caesar's cipher is cipher "B," and Julius Caesar's is cipher "D."

Caesarian ciphers are readily solved. *E*, for instance, might always be encoded as *U*. Then, since *E* is the commonest letter in many languages, *U* is likely to be the commonest letter in the ciphertext. That's a dead giveaway. A decipherer can identify the few most common letters, use them to recognize common short words, and quickly recover the message.

Cryptography has come a long way since the days of the Caesars. No nation today would use such a transparent cipher. However, simple Caesarian ciphers can be used to construct an unbreakable cipher, one that is used by modern superpowers.

The trick is to vary the Caesarian cipher from letter to letter. You use one of the 26 Caesarian ciphers for the first letter, another for the second, still another for the third, etc.

On the one hand, this complicates things enormously. You need a "key" telling which Caesarian cipher to use with which letter. The key must be at least as long as the message. The advantage is that the cipher is quite secure. Any letter of the ciphertext can encode anything at all. With a certain key, the Gettysburg Address would be encoded as a long string of *i*'s. With another key, it would be encoded as part of *Gulliver's Travels.* With another key (most possible keys, in fact), it would be the expected "random" mix of letters.

This type of cipher is the "one-time pad." The key is printed on a paper tablet. Each leaf of the pad is a key to be used once and then destroyed. The keys tell (for instance) which Caesarian ciphers to use with successive letters of the message. When the Caesarian ciphers are designated with letters as above, the keys look like random blocks of letters.

Use Caesarian ciphers "C," "R," "F," "B," "Z," "F," and "D" to encode the letters of the word *MESSAGE.* In cipher "C," *M* becomes *O.* In cipher "R," which looks like this—

A B C D E F G H I J K L M N O P Q R S T U V W X Y Z

R S T U V W X Y Z A B C D E F G H I J K L M N O P Q

—*E* becomes *V. MESSAGE* becomes *OVXTZLH.*

OVXTZLH is a much better encoding of *MESSAGE* than *NFTTBHF* is. When any one substitution cipher is used exclusively, clues remain that allow easy deciphering of any quantity of text. The entropy of the text can help identify the original language. From *NFTTBHF,* you know that the original word has a doubled letter in the middle and that the second and last letters are the same. Even on the basis of this single word, you might guess (correctly) that *F* stands for *E,* the commonest letter in English. Nothing in *OVXTZLH* is any help. The doubled *S* has become two different letters. Since a different, randomly selected substitution cipher was used for each letter, it is clear that *OVXTZLH* could just as easily stand for *any* seven-letter word at all. By being completely ambiguous, the one-time pad system ensures that any attempt at decoding without a key is pointless.

The problem with the one-time pad system is keeping the sender and receiver supplied with keys. Keys cannot be sent with the message. If they were, anyone intercepting the message could decipher it. Hence the pad. Rudolf Abel, a Soviet agent arrested in New York in 1957, had a one-time pad that was a tablet the size of a postage stamp, each page covered with fine print. An actual pad must have hundreds of numbers or letters to the page to accommodate messages of practical length. Such problems restrict the use of the one-time pad to important messages, and ones that are not very long.

With the right key, a one-time pad system will transform any text into *iiiii* . . . However, you have to tailor the key to a preexisting message. That defeats the usual purpose of a cipher, which is to communicate unknown future messages. Given that the ciphertext is going to be *iiiii* . . . no matter what, the system cannot communicate anything new or unexpected. The ciphertext just tells how much of the message to read off.

A ciphertext of *iiiii* . . . has the minimum possible entropy, far less than any real language. Usually, a cipher has as much or more entropy than the plaintext. When the entropy is less—as in the Voynich manuscript, unless it was originally in Tahitian!—it means that part of the message information has been shunted into the ciphering system. The ciphertext is ambiguous. Deciphering it depends on information not in the ciphertext: a key or the ability of the original author to reconstruct the message from an equivocal ciphertext.

Brute Force

Suppose that the Voynich plaintext is in a European language using the Roman alphabet, and that each of the manuscript's squiggles encodes exactly one letter in an unbreakable and unambiguous one-time pad scheme. Identify each of the Voynich symbols with a letter as Bennett did. Assume there are 26 distinct and meaningful symbols in all. With any particular "alphabetic" order for the symbols, there are 26 Caesarian ciphers encoding Roman letters into Voynich symbols.

We do not know the alphabetic order, if any, of the Voynich symbols. Nor do we know that the cipher system uses Caesarian ciphers only. There are many other ways of matching up letters to symbols. For the sake of simplicity, though, assume that the alphabetic order is known and that only Caesarian ciphers figure in the cipher system.

Then a key would specify which of 26 Caesarian ciphers was used for each symbol in the manuscript. If you had an alleged key to the manuscript's cipher, you could apply it to a few dozen symbols and see if the result was sensible words of some European language. If it was, you could apply it to the whole manuscript. If it produced a sensible message, the cipher would be solved.

We don't have a key. It might still seem that we could solve the cipher by brute force. We could check every possible decipherment of the Voynich text.

That is doubly impossible. It would entail checking all 26 keys for the first letter, and all 26 keys for the second, and all 26 keys for the third . . . (This isn't even the half of it, for it doesn't take into account the number of possible alphabetic orders of the Voynich symbols and the possibility of non-Caesarian ciphers.) This is 26^n, where n is the number of symbols in the sample of ciphertext. If a sample of 100 symbols is taken, the number of possible keys is 26^{100}. This comes to about 10^{141} keys—way too many to check, even if you had all the time in the world.

Yes, yes, but one can nevertheless imagine the hyper-herculean task of checking 10^{141} keys. It could be done in principle, however hopeless it might be in practice. *Still* the task is futile. Since we are checking every possible key, we are sure to come across a key that, applied in reverse, will transform the Voynich manuscript into *Gulliver's Travels*. For instance, take the first letter in *Gulliver's Travels*. One of the 26 Caesarian ciphers will transform that letter into the first symbol in the Voynich manuscript. Another Caesarian cipher will map the second letter into the second Voynich symbol, and so on. (There will be symbols left over at the end of the longer work.)

A different key will transform the Voynich manuscript into the Gettysburg Address. Other keys will decipher the manuscript into every possible text of an equal number of symbols. Even were it physically possible, an exhaustive search would be pointless because it would successively turn up every possible sensible message. All are equally inherent in the ciphertext.

Justifying a Decipherment

How, then, does anyone ever decipher anything (and more to the point, convince themselves and others that a decipherment is the correct one)? Consider this:

It is a fact verifiable by experience that choosing keys at random *never* produces a sensible decipherment. The chances against it are too great. Testing a random key on the Voynich manuscript invariably produces only a nonsense string of letters.

Consequently, the chance of an arbitrary *wrong* cipher producing a sensible decipherment is astronomically small. A key that produces a sensible message from a cipher is virtually certain to be the correct one *unless* the decipherer tailored the key to produce a desired message.

A convincing demonstration of the correctness of a decipherment has four steps:

First, you specify the cipher system and its key. Here "key" can mean whatever minimum amount of information must be remembered in order to apply the cipher, whether it takes the form of a printed key or not.

Second, you reverse the cipher system, producing an alleged plaintext from the ciphertext.

Third, the resulting plaintext is a sensible message, not nonsense.

Fourth, the key can be specified concisely. Examples of concise descriptions of keys are: "Use Caesarian cipher 'J' throughout"; "The key is the block of letters on a slip of paper found among Roger Bacon's effects"; "The key is the letters of the first page of the original edition of *Gulliver's Travels.*"

This fourth requirement is necessary to prevent decipherers from working backward from a hoped-for plaintext. The key must be "special." It must be intrinsically simple, or it must have historic justification. Most of the possible cipher keys have never been explicitly conceived by the mind of man. The ciphers that have been used, or even conceived, are in a very select minority of all possible ciphers. The decipherer must provide reason to believe that the key existed before he started working on the cipher.

The simplest cipher is the one that leaves any plaintext unchanged. Next simplest are those in which the same letter-substitution scheme is used throughout. The cryptograms found in puzzle books, in Edgar Allan Poe's "The Gold Bug," or Arthur Conan

Doyle's "The Adventure of the Dancing Men" are of this type. All such ciphers are in a very small minority, and a demonstration that a ciphertext may be converted to a sensible message via such a cipher is convincing evidence that the cipher was used and the message is genuine.

More difficult ciphers have a varying key that may be represented by a string of arbitrary letters (or numbers). If using the Gettysburg Address *as a key* converts a ciphertext into a sensible message, that is also a convincing demonstration that the message is genuine. (Keys taken from books or other texts are fairly common in real cryptography.) Although many, many keys will convert a given ciphertext to a sensible message, the chance of such a key being itself a sensible message is fantastically small.

This is not to say that the only valid complex keys are those that can be transliterated as literary passages. In a one-time pad, the key is random (else there would be no need for the pad). The random key on a one-time pad is special in a historical sense. Out of all the inconceivable number of possible keys, one was selected and printed on a pad. Finding a page from a one-time pad, and showing that it converts a ciphertext to a sensible message, is a convincing demonstration of the validity of the decipherment.

Where Is Meaning?

Where, then, is meaning: in the message, in the "key," or in the minds of those who understand it? Few would dispute that meaning is ultimately in the mind. It is in the mind the way that color or sound is. The more subtle question is whether the objective counterpart of the meaning resides more in the message or in the language or cipher system.

The answer is that it varies. A cipher consisting solely of the repeated letter *i* is an example of a system in which all the meaning is in the key. More often the meaning is split between message and key. It is, however, very hard to think of a case where *all* the meaning is in the message and none in the key. Ideally, this would be the case with a perfectly transparent language whose meaning is obvious to all (like airport pictographs, Esperanto, or artificial languages proposed for radio communication with extraterrestrial beings). None of the attempts at this come very close to that goal.

Scientific theories are expected to make sense of the world much as cipher keys do. Some theories put a lot of information in the theory; others only refer to information in the world.

The first extreme, the counterpart of Poe's *iiiii* . . . cipher, is exemplified by brains in vats. The brains-in-vats hypothesis requires an ad hoc assumption for everything. It rained yesterday because of a certain pattern of electrical stimulation. Roses are red because of a jolt of electricity; Gerald Ford became President because of another zap; moods, weather, animals, people, luck, and everything else are explained by stimulation. Brains in vats has no predictive power (compare the *iiiii* . . . cipher's inability to cope with unknown future messages). If we are brains in vats, the next apple you see could just as easily fall up as down. Anything could happen; you never know what those evil geniuses will come up with next.

At the opposite extreme is a theory like Newton's gravitation that points up an inherent regularity in the world. That apples don't fall up is built into the world, not into a series of ad hoc assumptions. The theory is simple, and it has predictive power.

Here as always it is impossible to say that one theory is indubitably right and the other wrong. It boils down to convenience. It is easier to use and remember the simpler hypothesis.

11

MIND

Searle's
Chinese Room

OF ALL THE MYSTERIES of the world, none is more puzzling than mind. Brain is a trifle in comparison. It is one thing to say that a mass of jelly, shaped by natural selection, can perform a wealth of complex functions. But where does consciousness come in?

Biology is well on the road to an understanding of how the brain works. Some say an understanding of consciousness is as remote as ever. The question of mind, long a favorite of philosophers, has lately become almost topical with advances in neurology, cognitive science, and artificial intelligence. The current state of thinking on mind is sharply illuminated by a set of intriguing paradoxes.

The Thinking Machine

The prisoner of the previous chapter's binary Plato's cave confronted sensory experience in the most abstract form. The irony is that our brains are such prisoners; the cave is the skull. It seems as if it would be impossible for anyone to duplicate how the brain deals with sensory information. This is the starting point for the thought experiments of this chapter.

The oldest picture of the mind is no picture at all. The caged parakeet that thinks its mirror image is another parakeet has no need of "mind" in formulating a worldview. This is not to say that the parakeet is stupid but only that it has no knowledge of self. The parakeet is aware of its bell and cuttlebone and the other objects of its world. This awareness may go so far as to predict the behavior of animate objects, such as the owner who fills the seed dispenser every morning. Is there anything the owner might do that would prove to the parakeet that he has a mind? No. The parakeet (a very intelligent parakeet, anyway) could ascribe any observed behavior to known and unknown causes and have no need to believe in mind.

It is worth noting that some extreme philosophical skeptics have espoused almost this view (note Hume's skepticism of his own mind). What, then, makes us think that other people have minds such as our own? A big part of the answer is language. The more we communicate with others, the more we are led to believe they have minds.

A second way of thinking about mind is dualism. This is the belief that mind or spirit or consciousness is something distinct from matter. We all talk this way, whether we believe in dualism or not: someone is full of spirit; they give up the ghost; not enough money to keep body and soul together. Dualism is motivated by the realization that there are other minds and that minds more or less stick to bodies.

As biologists have learned more about the human body, they have been impressed with the fact that it is made out of substances not so different from nonliving matter. The body is mostly water. "Organic" compounds can be synthesized. Physical forces such as osmotic pressure and electrical conductivity work in cells and account for much of their function. Mechanistic models of the body and brain have been so successful in limited ways that it is extremely attractive to think that they might account for all the myriad workings of the brain. This, a third way of thinking about the mind, supposes that the brain is a "machine" or "computer" of sorts, and that consciousness is—somehow—a result of that machine's operation.

Despite the modern trappings, mechanistic explanations for consciousness—and skepticism about them—are quite old. Gottfried Leibniz's "thinking machine," which he discussed in 1714, is equally cogent today:

> Moreover, it must be avowed that *perception* and what depends upon it *cannot possibly be explained by mechanical reasons,* that is, by figure and movement. Supposing that there be a machine, the structure of which produces thinking, feeling, and perceiving; imagine this machine enlarged but preserving the same proportions, so that you might enter it as if it were a mill. This being supposed, you might enter its inside; but what would you observe there? Nothing but parts which push and move each other, and never anything that could explain perception.

Leibniz's example is not especially compelling, but it does capture the malaise most of us feel about the mechanistic model. The thinking machine thinks, all right, but looking inside we find it as empty as a magician's trick cabinet. What did you expect to see?

A concise counterargument to Leibniz is David Cole's. Blow up a tiny drop of water until it is the size of a mill. Now the H_2O molecules are as big as the plastic models of H_2O molecules in a chemistry class. You could walk through the water droplet and never see anything *wet.*

The Paradox of Functionalism

Other thought experiments challenging the mechanistic model are much less easily refuted. One is Lawrence Davis's "paradox of functionalism."

Functionalism holds that a computer program that can do the same thing as a human brain must be comparable in every important respect, including having consciousness. The human brain can be idealized as a "black box" that receives sensory inputs from the nerve cells, manipulates this information in a certain way, and sends out impulses to the muscles. (Each vat in the brains-in-vats lab has two cables, one labeled "in" and one labeled "out.") What if there was a computer that, given the same inputs, would always produce the same outputs as a human brain? Would that computer be conscious? It is like Einstein and Infeld's sealed watch; one never really knows. Functionalism, however, says that it is most reasonable to believe that the computer would be conscious to the extent that that term has any objective meaning at all. One should believe this for the same reason one believes other people have minds: because of the way they act.

Davis proposed his paradox in an unpublished paper delivered at a conference in 1974. It deserves more attention than it has received. Suppose, he says, we learned about the sensation of pain in all relevant detail. Then (if the functionalists are right) we could build a giant robot capable of feeling pain. Like Leibniz's thinking machine, it is a really huge robot you can walk inside. The inside of the robot's head looks just like a big office building. Instead of integrated circuits, there are people in suits sitting behind desks. Each desk has a telephone with several lines, and the phone network simulates the connections of the neurons in a brain capable of feeling pain. Each person has been trained to duplicate the function of a neuron. It is boring work, but they get a competitive salary and benefits package.

Suppose that right now the set of phone calls among the bureaucrats is that which has been identified with excruciating pain. The robot is in agony, according to functionalism. But where is the pain? You won't find it on a tour of the office. All you will see is placid and disinterested middle management, sipping coffee and talking on the phone.

And the next time the robot is feeling unbearable pain, you visit and find that the people are holding the company Christmas party. Everybody's having a real ball.

The Turing Test

I will defer comment on Davis's paradox and go on to a closely related thought experiment, John Searle's "Chinese room." One further bit of background is necessary to appreciate the latter.

This is the "Turing test" of Alan Turing. In a 1950 essay, Turing asked whether computers can think. Turing argued that the question was meaningless unless one could point to something that a thinking agent can do that a nonthinking agent cannot. What would that difference be?

Already computers were performing calculations that had previously required the work of dedicated and intelligent human beings. Turing realized that the test would have to be something rather subtler than, say, playing a decent game of chess. Computers would soon do that, long before they would come close to being able to "think." Turing proposed as a test what he called the "imitation game."

A person sits at a computer terminal and directs questions to two beings, A and B, who are concealed in other rooms. One of the beings is a person, and the other is a sophisticated computer program alleged to be capable of thought. The questioner's goal is to tell which is the human and which is the computer. Meanwhile both human and computer are trying their best to convince the questioner that they are human. It is like a television panel show where the point is to distinguish an unknown person from an impostor.

The fact that the questioner communicates only via the computer terminal prevents him from using anything but the actual text of replies. He cannot expect to discern mechanical-sounding synthesized speech or other irrelevant giveaways. The concealed human is allowed to say things like "Hey, I'm the human!" This perhaps would do little good, for the computer is allowed to say the same thing. The computer does not have to own up to being the computer, even when asked directly. Both parties are allowed to lie if and when they think it suits their purpose. Should the questioner ask for "personal" data like A's mother's maiden name or B's shoe size, the computer is allowed to fabricate this out of whole cloth.

To "pass" this test, a computer program would have to be able to give such human responses that it is chosen as the human about half the time the game is played. If a computer could pass this test, Turing said, then it would indeed exhibit intelligence, *insofar* as intelligence is definable by external actions and reactions. This is no small claim.

That said, could a computer *think?* Turing concluded that the original question of whether computers can think was "too meaningless to to deserve discussion. Nevertheless, I believe that at the end of the century the use of words and general educated opinion will have altered so much that one will be able to speak of machines thinking without expecting to be contradicted."

In the years since Turing's essay, it has become common in cognitive science to associate mental processes with algorithms. If you execute a certain algorithm for calculating the digits of pi, then some small part of your thought process is directly comparable to the action of a computer calculating pi via the same algorithm. A widespread and popular speculation is that intelligence and even consciousness are like computer programs that may "run" on different types of "hardware," including the biological hardware of the brain. The functions of the neurons in your brain and their states and interconnections could in principle be exactly modeled by a marvelously complex computer program. If that program were run, even on a computer of microchips and wires, it would perhaps exhibit the same intelligence and even consciousness as you do.

Mind has long been thought of as soul, as *élan vital,* as half of the Cartesian dualism. Much of the intellectual community has abandoned these models in favor of a mechanistic picture of consciousness. John Searle's 1980 thought experiment caricatures the diminishing abode of mind into a shell game. If consciousness is nothing but algorithm, where does mind come in? Searle lifts the last shell and shows it to be empty.

The Chinese Room

Imagine that you are confined to a locked room. The room is virtually bare. There is a thick book in the room with the unpromising title *What to Do If They Shove Chinese Writing Under the Door.*

One day a sheet of paper bearing Chinese script is shoved underneath the locked door. To you, who know nothing of Chinese, it contains meaningless symbols, nothing more. You are by now desperate for ways to pass the time. So you consult *What to Do If They Shove Chinese Writing Under the Door.* It describes a tedious and elaborate solitaire pastime you can "play" with the Chinese characters on the sheet. You are supposed to scan the text for certain Chinese characters and keep track of their occurrences according to complicated rules outlined in the book. It all seems pointless, but having nothing else to do, you follow the instructions.

The next day, you receive another sheet of paper with more Chinese writing on it. This very contingency is covered in the book too. The book has further instructions for correlating and manipulating the Chinese symbols on the second sheet, and combining this information with your work from the first sheet. The book ends with instructions to copy certain Chinese symbols (some from the paper, some from the book) onto a fresh sheet of paper. Which symbols you copy depends, in a very complicated way, on your previous work. Then the book says to shove the new sheet under the door of your locked room. This you do.

Unknown to you, the first sheet of Chinese characters was a Chinese short story, and the second sheet was questions about the story, such as might be asked in a reading test. The sheet of characters you copied according to the instructions were (still unknown to you!) answers to the questions. You have been manipulating the characters via a very complex algorithm written in English. The algorithm simulates the way a speaker of Chinese thinks—or at least the way a Chinese speaker reads something, understands it, and answers questions about what he has read. The algorithm is so good that the "answers" you gave are indistinguishable from those that a native speaker of Chinese would give, having read the same story and been asked the same questions.

The people who built the room claim that it contains a trained pig that can understand Chinese. They take the room to county fairs and let people on the outside submit a story in Chinese and a set of questions based on the story. The people on the outside are skeptical of the pig story. The answers are so consistently "human" that everyone figures there is really a Chinese-speaking person in there. As long as the room remains sealed, nothing will dissuade the outsiders from this hypothesis.

Searle's point is this: Do you understand Chinese? Of course not! Being able to follow complex English directions is not the same as knowing Chinese. You do not know, and have not deduced, the meaning of a single Chinese character. The book of instructions is emphatically *not* a crash course in Chinese. It has taught you nothing. It is pure rote, and never does it divulge why you do something or what a given character means.

To you, it is all merely a pastime. You take symbols from the Chinese sheets and copy them onto blank sheets in accordance with the rules. It is as if you were playing solitaire and moving a red jack onto a black queen according to a card game's rules. If, in solitaire, someone asked what a card "meant," you would say it didn't mean anything. Oh, sure, playing cards once had symbolic significance, but you would insist that none of that symbolism enters into the context of the game. A card is called a seven of diamonds just to distinguish it from the other cards and to simplify application of the game's rules.

If you as a human can run through the Chinese algorithm and still not understand Chinese (much less experience the consciousness of a Chinese speaker), it seems all the more ridiculous to think that a machine could run through an algorithm and experience consciousness. Therefore, claims Searle, consciousness is not an algorithm.

Brains and Milk

Searle's skepticism is considerably more generous than many who doubt that computers can think. His thought experiment postulates a working artificial intelligence algorithm. It is the set of instructions for manipulating the Chinese characters. Clearly the algorithm must encapsulate much, much more than Chinese grammar. It cannot be much less than a complete simulation of human thought processes, and it must contain the common knowledge expected of any human to boot.

The story can be *any* story, and the questions can demand any fact, conjecture, interpretation, or opinion about it. The questions are not (at least, need not be) multiple-choice or questions asking you to regurgitate or complete lines from the story. Searle gave as an example this brief short story: "A man went into a restaurant and ordered a hamburger. When the hamburger arrived it was burned to a crisp, and the man stormed out of the restaurant angrily, without paying for the hamburger or leaving a tip." One question is: "Did the man eat the hamburger?" Now, the story doesn't say, nor will the Chinese character for "eat" even appear in the story. But anyone who understands the story will surmise that the man didn't eat the hamburger.

Questions could ask if a Big Mac is a hamburger (no way of telling from the story; you just have to know it) or if the story made you sad (the word/character for "sad" not appearing); they could ask you to point out sentences that made you laugh or to write another story based on the same characters. The algorithm must interact with the story much as a human would. Were this algorithm written in a computer language like LISP or Prolog, it would pass the Turing test. Searle avoids the "black box" mystery of a computer executing a sophisticated algorithm by dropping it in a human's lap.

Searle felt that the Turing test may not be all that it's cracked up to be. A computer that can act just like a human would be a remarkable thing whether "conscious" or not. We face the "other minds" problem in a more trenchant form. Even the skeptic does not doubt, in his nonphilosophical moments, that other people have minds. But we all have doubts about whether a machine may have a consciousness similar to ours.

Searle's opinion of the matter is surprising. He believed that the brain *is* a machine of sorts, but that consciousness has something to do with the biochemical and neurological makeup of the brain. A computer of wires and integrated circuits, even one that exactly duplicated the function of all the neurons in a human brain, would not experience consciousness (though it would function just like the human brain and pass the Turing test). A Frankensteinian brain—made "from scratch" out of the same kinds of chemicals as real brains—might be capable of consciousness.

Searle compared artificial intelligence to a computer simulation of photosynthesis. A computer program might well simulate photosynthesis in full detail (say, by creating a realistic animation of chlorophyll atoms and photons on a display screen). Though the program contains all the relevant information, it would never produce real sugar like living plants do. Searle felt that consciousness was a biological by-product like sugar or milk.

Few philosophers agree with Searle on this point, but his thought experiment has generated debate as few others have. Let's look at some of the reactions.

Reactions

One is that the experiment is flatly impossible. There can be no such book as *What to Do If They Shove Chinese Writing Under the Door.* The way we interpret language and think cannot be expressed in cut-and-dried fashion; it is something we can never grasp well enough to put it in a book. (Possibly, Berry's paradox and Putnam's Twin Earth lend credence to this position.) Therefore the algorithm *won't work.* The "answers" will be nonsense or the stock phrases of a talking Chinese teddy bear. They won't fool anyone.

This position is fine as far as it goes and cannot be refuted until the time, if any, that we have a working algorithm. Note only that Searle himself was willing to concede the possibility of the algorithm. And it is not strictly necessary to suppose that we will ever know how the brain *as a whole* works to conduct the experiment or one like it. You could implement Davis's bureaucratic simulation. The human brain contains approximately 100 billion neurons. As far as we know, the function of each individual neuron is relatively simple. The neuron waits for firing (electrical impulses) at its synapses, and when that firing meets certain logical criteria, it transmits an impulse. Suppose that we determine the exact state of one person's brain: the current states of all the neurons, the connections between them, and how each neuron works. Then all the world's population might participate in an experiment to simulate that person's brain. Each of the world's 5 billion people would have to handle the actions of about twenty neurons. For every connection between neurons, there would be a string between the two persons representing the neurons. Tugging on the strings would represent neural firing. Each person would operate the strings just as the neurons they represent would respond to neural firing. Again, however marvelous the simulation, no one would have any idea of what "thoughts" were being represented.

A second reaction is to agree with Searle that the algorithm would work but there would be no consciousness as a Chinese speaker. Searle's supporters cite the distinction between syntactic and semantic understanding. The rules in effect give the human syntactic understanding of Chinese, but no semantic understanding. He would not know that a certain character meant house and another meant water. Apparently, semantic understanding is essential for consciousness, and this is something computers can never have.

Nearly all those who disagree with Searle contend that there is some sort of consciousness buzzing around the Chinese room. It may be potential, incipient, slowed down, or brain-damaged, but it is there.

Chinese the Hard Way

Among the positions claiming a consciousness of Chinese, the simplest is that the subject would (contrary to Searle's claim) learn Chinese after all. There is a continuum between syntactic and semantic understanding. Maybe the rules would become second nature if applied long enough. Maybe the subject would surmise the meaning of the symbols from the way they were manipulated.

The heart of the issue is whether it is ever necessary to be told that "water" means *this,* and "house" means *this.* Or can we gather what *all* words mean from how they are used? Even if you had never seen a zebra, that does not prevent you from having semantic understanding of the word "zebra." You have certainly not seen a unicorn, and still have semantic understanding of the term.

Could you still have this semantic understanding if you have never seen a horse? Never seen an animal of any kind (not even a human)? At some degree of isolation from the object, you must wonder whether understanding would exist.

Suppose you were sick and missed the first day of arithmetic class, the day when the teacher explained what numbers are. When you return to school, you're afraid to ask what numbers are because everyone else seems to know. You make an extra effort to learn all the subsequent material, like the addition tables, fractions, and so on. You make such an effort that you end up being the top student in math. But inside you feel you're a fraud—you still don't know what numbers are. You only know how numbers *work,* how they interrelate with each other and everything else in the world!

One feels that that is all the understanding that anyone can have of numbers (though possibly numbers differ from zebras in this respect). A similar case is Euclidean geometry. The study of geometry typically starts with the disclaimer that concepts such as "points" and "lines" will not be defined as such and should be taken to mean only what may be inferred from the axioms and theorems about them.

An objection to this position is that the human starts cranking out Chinese answers right away—before he could have memorized the rules or deduced the meanings of characters. For a long time, the questioners outside the room will be able to pose questions whose answers will be new words that the subject has never used before. ("What's that condiment some people put on hamburgers, made from chopped-up pickle?" Will Searle's subject be able to deduce the meaning of the character for "chowchow"?)

Dr. Jekyll and Mr. Hyde

Some claim that the human simulator would understand Chinese but not know it. David Cole compared Searle's subject to a bilingual person with a peculiar type of brain damage that prevents him from being able to translate. Or (take your pick) he would be like a person with multiple personalities, a "split brain" patient, or an amnesiac.

Dr. Jekyll goes into the room, speaking only English. Running through the algorithm creates a Mr. Hyde who speaks Chinese. Jekyll doesn't know about Hyde, and vice versa. Consequently the subject is unable to translate between English and Chinese. He is unaware of his Chinese ability, and even denies having it.

We have many mental capacities of which we are unaware. Right now your cerebellum is regulating your breathing, eye blinking, and other automatic functions. Normally these functions are on autopilot. You can take conscious control of them if desired. Other functions, like pulse, are more automatic and can be controlled consciously only to a degree, through biofeedback techniques. Still more automatic functions may not be controllable at all. All are under the guidance of your one brain.

If this is the case, why are the two linguistic personalities so poorly integrated? Maybe because of the bizarre way that a knowledge of Chinese has been "grafted" onto the subject's brain.

The Systems Reply

In his original article, Searle anticipated several responses to his thought experiment. He called one the "systems reply." This says that indeed the person would not "know" Chinese, but the process —of which the person is a part—might in principle. The person in Searle's Chinese room is not analogous to our mind; he is analogous to a small but important part of a brain.

The systems reply is no straw man. Broadly interpreted, it is the most popular resolution of the paradox among cognitive scientists. Not even the most dogmatic mechanist presumes that individual neurons experience consciousness. The consciousness is in the process, of which the neurons are mere agents. The person in the locked room—the book of instructions—the sheets of paper shoved under the door—the pen the person writes with—are agents.

Searle's counterargument to the systems reply went like this: All right, assume there *is* consciousness in the system composed of the human, the room, the books of instructions, the bits of scratch paper, the pencils, and anything else used. Tear down the walls of the room; let the human work alfresco. Have him memorize the instructions and do all the manipulations henceforth in his head. If the pencils are suspect, have him scratch out the answers with his fingernails. The system is reduced to just the human. *Now* does he understand Chinese? Of course not.

One danger with thought experiments is that their convenience can lead us astray. You must make sure that the reason you're just fantasizing rather than doing the experiment isn't something that would invalidate the fantasy. Most philosophers and scientists of the systems-reply camp feel that this is the essential problem with Searle's Chinese room.

A Page from the Instructions

It may be helpful to analyze the mechanics of the situation in slightly greater detail. Reverse the situation so that the human is a Chinese speaker and reader who knows nothing of English or the

Roman alphabet (this is more convenient here, for we will discuss how to understand English). Let the story be an English translation of Aesop's fable of the fox and the stork, and the next day's batch of writing poses questions about both animals. Consider what the text of a (Chinese-language) book entitled *What to Do If They Shove English Writing Under the Door* must be like.

Part of the discussion must tell you how to recognize the word "fox." We know that only words, rather than letters, have meaning in an English text. Therefore any algorithm that simulates reasoning about the characters and events of a story must isolate and recognize the words naming them. An English reader recognizes "fox" at a glance. Not so the Chinese reader. He must follow a laborious algorithm that may go something like this:

1. Scan the text for a symbol that looks like one of these symbols:

F f

If you find the symbol, go to step 2. If there is no such symbol in the text, go to the instructions on page 30,761,070,711.

2. If there is a blank space immediately to the right of the symbol, go back to step 1. If there is a symbol immediately to the right, compare it to these symbols:

O o

If the symbol matches, go to step 3. If not, go back to step 1.

3. If there is a blank space immediately to the right of the symbol in step 2, go back to step 1. If not, compare the symbol to these symbols:

X x

If it matches, go to step 4. If not, go back to step 1.

4. If there is a blank space or one of these symbols immediately to the right of the symbol in step 3, go to the instructions on page 84,387,299,277. If there is a different symbol to the right, go back to step 1.

•   ,   ;   :   ”   ’   !   ?

These instructions haven't gotten us anywhere yet. Who knows how much more complicated is the part of the instructions telling how to *think* about foxes?

We do not have Searle's algorithm for understanding Chinese, but we do have simpler ones. A very naïve person, who has never seen a pocket calculator before, might get the woefully wrong idea that it can think. You could disabuse him of this notion in a Searle experiment. Give him the specs for the microprocessor used in the calculator, and a wiring diagram, and specify the electrical inputs that would result from punching out a problem on the calculator's keys. Let him run through the action of the microprocessor as it does some math. The human-simulating-microprocessor would produce the correct results, but would have no idea of the actual mathematical operation being carried out. He wouldn't know whether he was adding 2 plus 2 or taking the hyperbolic cosine of 14.881 degrees. The person would feel no consciousness of the abstract mathematical operation, and neither would the pocket calculator. If anyone tried to argue the systems reply, you could have the subject memorize everything and do it in his head. Right?

Don't be so sure. A calculator may run through thousands of machine steps to perform a simple calculation. The experiment would probably take *hours*. Unless the subject had a phenomenal memory, it would be impossible for him to do the microprocessor simulation in his head. He would almost certainly forget some intermediate results along the way and ruin everything.

Now consider the plight of Searle's subject. The book of instructions must be very big indeed! It must be far, far bigger than any room on earth.

Since no one has devised an algorithm for manipulating Chinese characters able to "answer" questions intelligently, it is impossible to say how big or complex the algorithm would be. But given that the algorithm must simulate human intelligence, it is reasonable to think that it cannot be much less complex than the human brain.

It is conceivable that each of 100 billion neurons plays some part in actual or potential mental process. You might expect, then, that the instructions for manipulating Chinese symbols as a human does would have to involve at least 100 billion distinct instructions. If there is one instruction per page, that would mean 100 billion pages. So the "book" *What to Do If They Shove Chinese Writing Under the*

Door would more realistically be something like 100 million volumes of a thousand pages each. That's approximately a hundred times the amount of printed matter in the New York City library. This figure may be off by a few factors of 10, but it is evident that there is no way anyone could memorize the instructions. Nor could they avoid using scratch paper, or better, a massive filing system.

It's not just a matter of the algorithm *happening* to be impractically bulky. The Chinese algorithm encapsulates much of the human thought process, including a basic stock of common knowledge (such as how people act in restaurants). Can a human brain memorize something that is as complex as a human brain? Of course not. You cannot do it any more than you can eat something that is bigger than you are.

By the same token, you've probably seen statistics like "The average American eats an entire cow every six months." A cow is bigger than a person, but the statistical beef eater consumes a little bit of a cow at a time. There is never very much of a cow inside you at any time. So it would be with Searle's subject.

Since the brain is made of physical stuff and stores memories as chemical and electrical states of this physical stuff, it has a finite capacity to remember things. It is not clear how much of the brain is available for storing memories, but certainly not all of it, and perhaps only a small fraction. Other parts of the brain have to be available to manipulate memories, process new sensory information, etc.

Evidently, all the variations on the thought experiment (from Searle and his critics) that have the subject memorizing the rules are misleading. It is impossible for the person to memorize anything more than a tiny fraction of the complete algorithm. He must constantly refer to the instructions and to his scratch sheets/filing system. Frequently the instructions will refer him to a certain scratch sheet, and he will look at it and shake his head: "Gee! I can't even remember writing this!" Or he will turn to a page in the instructions, see a coffee ring, and know that he has consulted the page before but not remember it.

The human is, ultimately, a very small part of the total process. He is like a directory assistance operator who looks up thousands of phone numbers every day but cannot remember them a few moments after reciting them. The information about phone numbers is practically all in the phone books; and in Searle's experiment, the algorithm exists mostly in the instructions and scratch sheets and hardly at all in the human or the tiny fraction of the instructions he remembers at the moment.

That the human in the room is a conscious being is irrelevant and rather disingenuous. He could be replaced by a robot (not a slick, science-fictional robot with artificial intelligence; just a device, maybe a little more complicated than a mechanical fortune-teller). The fact that the human fails to experience a second consciousness is no more significant than that Volume 441,095 of the instructions does.

This explains the human's denial of understanding Chinese. It is less satisfying in saying where and how the consciousness exists in the process. We want to point to the scratch paper, instructions, and so on, and say, "The consciousness is right over there by that filing cabinet." About all we can do is to postulate that we are failing to see the forest for the trees. We are like the man inside Cole's giant water drop who can see nothing wet.

The Chinese room is dilated in time more than in space. Imagine we have a time machine that can accelerate the Chinese room a trillionfold or so. *Then* the pages of the instructions would be a blur. Stacks of scratch paper would appear to grow organically. The human, moving too fast to see, would be a ghost in the machine. Possibly, part of our concept of consciousness requires that things be happening too fast to keep up with them.

A Conversation with Einstein's Brain

Douglas Hofstadter devised a thought experiment (1981) in which the exact state of Einstein's brain at death is recorded in a book, along with instructions for simulating its operation. By carefully applying the instructions, you can have a (very slow) posthumous conversation with Einstein. The responses thus derived are exactly what Einstein would have said. You have to address the book as "Albert Einstein" and not a book, because it "thinks" it is Einstein!

Hofstadter's thought experiment neatly splits the presumed consciousness into information (the book) and process (the person following the book's instructions). Everything that makes the book Einstein is in the book. But the book, sitting on a shelf, patently has no more consciousness than any other book. This leads to an ingenious set of riddles on the "mortality" of Searle simulations.

Suppose that someone patiently applies the rules in the book at the rate of so many instructions per day. Einstein's consciousness is, or seems to be, re-created. After a while the human replaces the book on the shelf and takes a two-week vacation. Is the book "Einstein" dead?

Well, the book could no more "notice" the hiatus than we could detect it if time stopped. To the "Einstein" of the book, the person is analogous to the physical laws that keep our brains ticking.

What if the person carrying out the instructions slowed down to one instruction a year? Is that enough to keep the book "alive"? What about one instruction a century? What if the interval between instructions doubles each time?

12

OMNISCIENCE

Newcomb's Paradox

F EW CONCEPTS are more inherently paradoxical than omniscience. Most cultures believe in a superior being or beings with total knowledge. Yet omniscience readily leads to contradiction. In part the trouble is that there is something suspect about utter perfection in *anything*. At the very least, omniscience, if it exists, has some unexpected properties.

The most dazzling of paradoxes of omniscience is of recent vintage (1960). Devised by physicist William A. Newcomb, the paradox has spurred almost unprecedented interest in the scientific community. *(The Journal of Philosophy* called it "Newcombmania.") Besides exploring the issues of knowledge and prediction, Newcomb's paradox offers a new twist on that philosophical standard, free will.

Before getting to Newcomb's paradox, it will be instructive to approach it via two simpler but related situations from game theory, the abstract study of conflict.

The Paradox of Omniscience

The "paradox of omniscience" shows that being all-knowing can be to your disadvantage. The paradox is described in the context of a deadly diversion of game theorists and 1950s teenagers called "chicken." This is the adolescent dare game in which two drivers race toward each other in a collision course. You are in the driver's seat of a car traveling at high speed in the middle of a deserted highway. Your opponent is in an identical car, traveling at the same speed toward you. If neither of you veers to the side, both will crash and die. Neither of you wants that. What you *really* want is to show your machismo by not swerving—and having your opponent swerve (lest you both get clobbered). Failing that, there are two intermediate scenarios. It wouldn't be so bad if both you and your opponent chickened out. At least you'd survive, and wouldn't suffer the humiliation of being the one who flinched while his opponent kept his cool. Of course, even the latter would be better than instant death in a head-on collision.

In game theory, chicken is interesting because it is one of a few fundamental situations in which the best course is not immediately apparent. When the game is played between ordinary mortals, each driver's situation is identical. In the long run, the best either person can do is to chicken out, in the hope that the opponent will be smart enough to do the same. If one driver does *not* swerve, the opponent will be angry and may not swerve the next time, with dire consequences for both. In short, no chicken player reaches middle age except by being a consistent coward.

Now imagine playing chicken with an omniscient opponent. The other driver is gifted with infallible ESP. He can and does anticipate your moves with perfect accuracy. (You are still an ordinary mortal.) "Oh-oh!" you think. "The whole point of chicken is guessing what the other guy will do. I'm in big trouble!"

Then you mull over your predicament a bit and realize that you have an unbeatable advantage. It is foolish to swerve with an all-knowing opponent. He will predict your swerve and thus won't swerve himself—resulting in complete failure for yourself.

Your best course is not to swerve. Anticipating *that,* Mr. Know-it-all has only two options: to swerve and survive (albeit with humiliation) or not to swerve and die. Provided he is rational and does not want to die, he can only swerve. Consequently, *the omniscient player is at a disadvantage.*

The paradox of omniscience is merely of the "common sense is wrong" variety. Surprising as its conclusion may be, it is valid and not like the contestable reasoning of the prisoner in the unexpected hanging. The omniscient driver cannot negotiate his way out of the disadvantage either. Allow the drivers a tête-à-tête before the game. The omniscient driver can take one of two bargaining positions:

1. "Make my day." He can play tough by threatening to swerve if and only if you swerve.
2. "Look at the long term." He can appeal to your sense of wisdom (or your knowledge of game theory): "Sure, you might get away with not swerving this time. But look at the long term. The only course that works in the long run is for both of us to swerve."

The first strategy's threat has no teeth. The omniscient driver can bluster all he likes, but if he foresees that you *aren't* going to swerve, would he really *not* swerve and kill himself? Not if he isn't suicidal.[1] The second strategy, which appears to be 180 degrees removed from the first, falls victim to the same counterstrategy. You still need only resolve not to swerve to create a swerve-or-die situation for the omniscient driver.

Situations like chicken (and implicit paradoxes of omniscience) occur frequently in the Old Testament. Adam, Eve, Cain, Saul, and Moses challenged an omniscient God, who had told them that disobedience, while pleasurable in the short run, would be ruinous in the long run. The paradox is weakened by the fact that the omniscient deity is also all-powerful and can presumably overcome any disadvantage deriving from His omniscience.

Even today chicken is being played all the time. Game theorists suggested chicken as a metaphor for the 1962 Cuban missile crisis, with the United States and the Soviet Union as the players. In geopolitical contexts, the paradox of omniscience calls the value of espionage into question. An all-knowing nation may be at a disadvantage in some situations (note that the paradox does not say that omniscience is disadvantageous in *all* situations). For the paradox to apply, nation A must have such a vast network of spies that it

---

[1] This may suggest a third strategy: pretending to be suicidally despondent. If the omniscient being could convince you that he wanted to die, then you would have to swerve to save yourself. Clever though this is, it is not quite chicken (cricket?). As game theorists define chicken, the actual preferences of each player are known.

can learn of every high-level decision in nation B. Nation B must be aware that it is hopelessly riddled with moles and cannot keep a decision secret from nation A. (The nonomniscient player must always be aware of the opponent's omniscience for the paradox to apply.) Ironically, the latter requirement may prevent the paradox from occurring much in the real world: Few governments are willing to acknowledge their security leaks.

The Prisoner's Dilemma

Newcomb devised his paradox while pondering the "prisoner's dilemma," another well-known situation in game theory. It is worth describing briefly that situation too.

In the prisoner's dilemma, two wrongdoers are arrested for a crime. The police interrogate the prisoners separately so that they may not collaborate on a lie. Each prisoner is offered a deal. The corrupt police want a scapegoat. If the prisoner confesses everything, they'll let him go (provided the partner doesn't do the same thing). Each prisoner must make his decision without consulting the other, and knowing that the other is being offered the same deal. What is the best course for a prisoner to take?

The best individual outcome for either prisoner results when he confesses and his partner doesn't. That lets him off the hook completely. Conversely, the *worst* fate is to be the one who doesn't confess. Given the partner's testimony, the judge is sure to throw the book at the guilty party who persists in his lie.

Things are almost as bad if both confess. Then both will be convicted. Still, neither is quite as bad off as in the case where his partner gets off scot-free: The fury of the law is split between the two. Likewise, things are fairly good for both parties if both don't confess. The police will still suspect the pair, but they may not have enough evidence to secure a conviction.

The prisoner's dilemma explores the conflict between the good of the individual and the good of all. The prisoners really shouldn't confess, because that is best for the pair. But assuming the other guy won't confess, each prisoner is tempted to better his own situation by turning state's evidence. The real-world versions of this are so numerous and obvious they needn't be enumerated.

As you will gather, the prisoner's dilemma is closely related to chicken. In both the participants are tempted to do something that would be disastrous if both did it (not swerve, turn state's evidence). Call this course of action "defecting." In chicken the worst possible outcome results when both players defect. In the prisoner's dilemma, the worst outcome results when the other guy defects and you don't. Thus the temptation to defect is more acute in the prisoner's dilemma. In chicken, if you know your opponent is going to defect (by being omniscient, say), you can only grit your teeth and *not* defect. In the prisoner's dilemma, knowing your partner will defect is all the more reason for you to defect.

Newcomb's Paradox

Newcomb's paradox goes like this: A psychic claims to have the ability to predict your thoughts and actions days in advance. Like most psychics, he does not claim total accuracy. He is right about 90 percent of the time. You have agreed to take part in an unusual test of the psychic's powers. A TV news program has provided the facilities and put up a large sum of money; all you have to do is abide by the conditions of the experiment.

On a table in front of you are two boxes: A and B.

Box A contains a thousand-dollar bill. Box B either contains a million dollars or is empty. You cannot see inside it. Of your own free will (if there is such a thing), you must choose either to take box B only or to take both boxes. Those are the only options.

The catch is this: Twenty-four hours ago, the psychic predicted what you would choose. He decided whether to put the million dollars in box B. If he predicted that you would take *only* box B, he put the million dollars in it. If he foresaw your taking both boxes, he left box B empty.

Personally, you don't care whether the psychic's powers are proven or discredited. Your only motive is leaving the experiment with as much money as possible. You are not so wealthy that money means nothing. The thousand dollars in box A is a lot of money to you. The million dollars is a fortune.

The conditions of the test have been and will be enforced scrupulously. You need entertain no doubts that box A contains the $1000. Box B can contain only $1 million or nothing at all, based on the psychic's prediction. No one is trying to trick you on that score. A trusted friend was present at the time the psychic made his prediction, and made sure he obeyed the rules about putting the money in the boxes.

Just as certainly, *you* will be prevented from circumventing the rules. Armed guards will prevent nihilist acts like not taking *any* box. Nor can you cheat the psychic by basing your decision on something other than your own mental processes. You can't decide on the basis of a coin toss or whether the number of shares traded that day is odd or even. You have to analyze the situation and decide on the most profitable of the two options. Of course, the psychic has anticipated your analysis. What should you do—take both boxes or just B?

Reactions

One reaction to the paradox goes like this: Psychics! Everyone knows that's a lot of hooey! So all that business about the "prediction" is irrelevant. It boils down to this: There are two boxes, they might contain money, and you're free to take them.

It's silly to take just box B when there is a guaranteed $1000 in box A. That is exactly like not picking up a thousand-dollar bill you see on the street. The contents (if any) of box B are not going to vanish if you take both boxes. No one, including the psychic, claims that that sort of telekinetic power is at work. The boxes were hermetically sealed twenty-four hours ago. You should take both.

There is also a strong argument for taking only box B. Remember, the psychic *is* usually right. That is a given. Chances are, he would be right about your taking both boxes, in which case you'd get a measly thousand. Meanwhile, a sucker taken in by his claims would get a million.

What if the experiment has been conducted hundreds of times previously, with the psychic almost always being right? This should not change things, since the psychic's accuracy is postulated. Bookmakers are accepting side bets on the outcome of the experiment. Provided you take box B only, they quote odds of 9 to 1 in favor of its containing a million dollars. If you take both boxes, your chance of getting a million is a long shot at 9 to 1 against. The bookmakers are not quoting these odds out of altruism. These are the actual probabilities, as near as anyone can determine them.

Money being the only thing that matters in the experiment, the case for taking box B can be stated in dollars and cents. If you take both boxes, you will get a sure $1000 (box A) and further stand a 10 percent chance at a million dollars—the latter if the psychic wrongly predicted that you would take box B only. On the average, a 10 percent chance of $1 million is worth $100,000. The total expected profit from taking both boxes is therefore $1000 + $100,000 or $101,000.

If instead you take box B only, you stand a 90 percent chance that the psychic will be right and leave a million dollars in there. That's worth $900,000 on the average. It is strongly in your favor to take B only. The better the psychic's batting average, the more you profit by taking only box B. If he is right 99 percent of the time, the strategies would be worth $11,000 (both boxes) and $990,000 (B only). In the limiting case where he's *always* right, it amounts to a choice between $1000 (both boxes) or $1 million (B only).

No one has yet squared these opposing viewpoints to everyone's satisfaction. The ingenuity of suggested resolutions to Newcomb's paradox may be without parallel. Among the bizarre explanations that have been seriously offered are that the sealed boxes constitute a Schrödinger's cat–like situation and are neither empty nor full until opened!

The conventional analysis of the prisoner's dilemma is wanting. Notice the parallels. As with the prisoners, you and the psychic really should "cooperate" by predicting, and then taking, box B only. But assuming that the psychic did cooperate, you are sorely tempted to enrich yourself further by taking both boxes. It is a contention of game theory that one should never be the first to defect in a prisoner's dilemma situation. But how can that advice apply here? The psychic has played his hand, and there are no future consequences to worry about.

Glass Boxes

Many variations on the basic situation have been proposed in the attempt to make the correct course clearer. The predictor can be a being from another planet, God, a spouse of twenty years who "knows how you think," or a computer that has been programmed with extensive information on the state of all the neurons in your brain. You can vary the predictor's accuracy from 50 to 100 percent to see what difference it makes. Some variations stack the deck in favor of one choice, but none exorcises the paradox.

The paradox depends on faith in the predictor's abilities. Suppose the "psychic" has no predictive power whatsoever and just tosses a coin to decide whether to put the million in box B. In that case, everyone must agree that you should take both boxes. Whether the predictor is right or wrong, you're $1000 ahead for taking both boxes. Figuring the odds leads you to the same conclusion. Taking both boxes gets you a sure $1000 plus a 50 percent chance at $1 million ($501,000 total) vs. 50 percent of $1 million ($500,000) for taking just box B.

The paradox further requires that the predictor's accuracy be great enough to make up for the loss of the sure thing in box A. The accuracy must be greater than 50.05 percent with the dollar amounts stated. In general, the accuracy must be at least $(A + B)/2B$, where A is the amount in box A, and B is the amount that may or may not be in box B.

The argument for taking both boxes is more trenchant if box A is made of glass and box B has a glass window on its far side. You can verify for yourself that the $1000 is in box A. A nun who has taken a vow of truthfulness sits on the opposite side of the table and can see in box B's window. The nun is not allowed to betray the contents of box B by facial expressions or any other means, but after the experiment is over she will be able to affirm that the money didn't vanish or appear out of thin air as you made your choice. With this setup, wouldn't you feel silly just taking box B? The psychic has already committed himself. The nun will see you either passing up a sure $1000 and taking an empty box B—then you'd *really* feel stupid—or getting the million but still passing up the thousand dollars for no reason whatsoever.

Prior to the experiment, you announce that 10 percent of your proceeds will be donated to an orphanage. The nun, who sees what's in the boxes, silently prays that you will do whatever will result in the greater donation. *There can be no doubt what the nun wants you to do.* She wants you to take both boxes. No matter what she sees, your taking both boxes means an extra $100 for those needy waifs.

In still another variation, this one suggested by Newcomb, both boxes are glass on all sides. Box B contains a slip of paper on which is written a very large odd integer. The experiment's sponsors have agreed to pay the bearer of that slip of paper $1 million *if* the number is prime. The psychic has chosen the number so that it will be prime only if he predicted you would take only box B. You can see the number and make a record of it for reference, but you are not allowed to determine if it is prime until you have made your choice. Now, certainly a mathematical fact is not going to change. Long before the stars existed, there was number; nothing you do here and now on this insignificant planet will make any difference in the realm of mathematics. This version of the paradox is one last chance to cast out any doubts you may have about your decision affecting the prediction through some kind of weird backward causality.

Like an automobile's chronic knock, the paradox lingers, even when you start to disassemble it. Suppose that the experimenters elect to sweeten the odds in your favor. Under modified rules, you are allowed—in fact, encouraged—to open box B first and *then* decide if you want to take box A as well. After you open B and see what's in it, you can clutch the million (if it's there) tightly to your bosom, and even deposit it in your bank account if you still harbor some silly idea that the money may yet go *poof!* and vanish. Then, *only then,* must you decide whether to take the $1000 in box A as well.

Are we not all agreed that you would be stark, raving, foaming-at-the-mouth mad *not* to take A? You'd certainly take it if you found box B empty. It would be equally irrational not to take A after finding the million in B and banking it.

That agreed, not everyone is rational. There would still be an occasional moron who would open box B, find a million dollars, and not take box A. Of course, everyone with half a brain opens B, finds nothing, and takes box A.

Prediction of human behavior brings up questions of free will. You can excise free will from Newcomb's paradox thus: The psychic is not what he claims. He has no powers of clairvoyance. Instead, he has a device that causes the subjects to choose whichever option he selects. The psychic decides you'll take both boxes, fixes them accordingly, *and then pushes a button that makes you take both boxes.*

This eliminates concern that the prediction in Newcomb's paradox is physically impossible. Of course, we can no longer ask, "What would you do?" You'd do whatever the psychic decided you'd do. The most we can ask is, "Who would you rather be, one of the zombies who get $1000 or one of the zombies who get $1 million?" Well, of course, you'd rather get the million. Some *real* money is the least you deserve, giving up your free will like that.

If you agree with that, does it really matter *how* the psychic achieves his accuracy, through prediction or mind control? You're concerned only with money, not with making some existential statement. Shouldn't you then take box B even if there is no mind control and you do have free will?

Opinion remains strongly divided on Newcomb's paradox. There is this distinction between the both-boxes and the box-B camp: Everyone who would take box B only does so in the expectation of getting the $1 million. The both-boxes people are divided among those who austerely expect to get only the $1000 and those who imagine they have a greater than stated shot at the $1 million too.

If I was convinced that Newcomb's situation truly existed, I would take box B only. I'm not saying that's "right" but only that that's what I'd do. This seems to be the most popular choice, and is in keeping with the game-theoretic analysis of the prisoner's dilemma, for what that's worth. Newcomb felt you should take only box B. Many philosophers take the opposite position.

Nozick's Two Principles of Choice

One of the most insightful analyses of the paradox is Robert Nozick's "Newcomb's Problem and Two Principles of Choice," published in *Essays in Honor of Carl G. Hempel* (1969). Nozick pointed out that the paradox calls two time-tested principles of game theory into conflict. One principle is that of *dominance*. If a certain strategy is always better than another, whatever the circumstances, then it is said to dominate the other strategy and should be preferred over it. Here the strategy of taking both boxes dominates the strategy of taking B only. No matter what the psychic did, you are $1000 the richer for taking both boxes.

Just as unquestioned is the principle of *expected utility*. It says that if you total the gains from alternative strategies (as done above), you should choose the one with a greater expected gain. No one had expected that these two principles could be in conflict.

It's not that simple, though. Whether a strategy dominates another can depend on how you look at the situation. Suppose you must choose between betting on either of two horses, Sea Biscuit and Hard Tack. It costs $5 to bet on Sea Biscuit, and you win $50 (plus the return of the original $5) if he wins. It costs $6 to bet on Hard Tack, and you are ahead $49 if he wins. This can be summarized in a table:

	SEA BISCUIT WINS	HARD TACK WINS
Bet on Sea Biscuit:	Win $50	Lose $5
Bet on Hard Tack:	Lose $6	Win $49

What should you do here? Neither of the two permissible wagers dominates. Obviously, it's better to bet on Sea Biscuit if Sea Biscuit wins, and on Hard Tack if Hard Tack wins. In this case you must use the expected utility principle, which looks at the probabilities of the two horses winning. Suppose that Hard Tack actually has a 90 percent chance of winning and Sea Biscuit only a 10 percent chance. Then you would certainly want to bet on Hard Tack.

Now look at things a little differently. Instead of categorizing the possible states of affairs by the horse that wins, talk about your luck. Consider your gains or losses if you are lucky in your bet or unlucky in your bet:

	YOUR HORSE WINS	YOUR HORSE LOSES
Bet on Sea Biscuit:	Win $50	Lose $5
Bet on Hard Tack:	Win $49	Lose $6

Now betting on Sea Biscuit *does* dominate betting on Hard Tack. If your horse wins, you are ahead a dollar, and if your horse loses, your loss is a dollar less.

Something is peculiar here. Both tables accurately describe the payoffs. The difference may suggest that between Goodman's "grue" and "green." But the two ways of categorizing the situation (by the *name* of the winning horse and by whether *your* horse wins or loses) are natural ways of talking, not made-up categories like "grue" and "bleen."

The conflict, Nozick surmised, comes from the fact that the second states (your horse wins/your horse loses) are not "probabilistically independent" of your decision. Your choice of which horse to bet on influences the chance of being lucky or unlucky. Sea Biscuit is the long shot. Bet on him, and chances are that you will be unlucky. Betting on the shoo-in, Hard Tack, raises the odds that you will be lucky.

From this Nozick concluded that it is valid to use the dominance principle only when one's choice does not affect the outcome. Try out this rule in the paradox. The dominance principle, which tells you to take both boxes, is unreliable if your choice can influence the psychic's prediction. That would be possible only if there was backward causality. This is generally presumed impossible. This rule fails to resolve the paradox.

Nozick then considered other intriguing scenarios. It is possible that one's choice has no causal effect on an outcome but is nonetheless probabilistically linked to it.

What about a hypochondriac who has memorized the symptoms of all known diseases and reasons thus: "I'm a little thirsty; I think I'll have a glass of water. I've sure been drinking a lot of fluids lately. Oh-oh! Excessive thirst is a symptom of diabetes insipidus. Do I *really* want that glass of water? Guess not."

Everyone agrees that this is ridiculous. Drinking water does not cause diabetes. It is the height of absurdity to base a choice on whether to have a glass of water on its pathological correlations. This is not to say that the pathological correlations aren't legitimate. A desire for water *is* (very slight) confirmation for a hypothesis that one has a disease whose symptoms include a craving for water. The fallacy is basing a choice on the correlations. The hypochondriac is (literally) treating the symptoms, not the disease.

Nozick compared Newcomb's situation to a prisoner's dilemma with two identical twins. A prisoner and his identical twin are being held incommunicado, each independently considering whether to turn state's evidence. Suppose, Nozick said, it has been established that behavior in prisoner's dilemma situations is genetically determined. Some people's genes cause them to cooperate; others are congenitally inclined to defect. Environment and other factors enter into it too, but say that one's choice is 90 percent determined by the genes. Neither prisoner knows which gene he and his twin have. Each prisoner might reason like this: If I defect, chances are my twin brother will defect too, having identical genes. That will be bad for both of us. If I cooperate, my twin probably will as well—which isn't a bad outcome at all. So I should cooperate (with the twin; refuse to turn state's evidence).

The diagram looks like this. The outcomes are expressed for both twins in arbitrary units. "(0,10)" means the worst possible outcome for twin 1 and the best possible outcome for twin 2. The two genetically favored (?!?) outcomes, where both twins act identically, are italicized.

	TWIN 2 TURNS STATE'S EVIDENCE	TWIN 2 REFUSES TO TALK
TWIN 1 TURNS STATE'S EVIDENCE	*1,1*	10,0
TWIN 1 REFUSES TO TALK	0,10	*5,5*

Is not this reasoning just as silly as the hypochondriac's? Twin 1's choice cannot affect twin 2's decision, much less "reach back" and affect their genes. Either the twins have the gene or they don't. Although cooperating may not be such a bad idea, it is unreasonable to use the genetic correlation to make the decision.

Nozick's essay ends by asking how the Newcomb situation is any different from the twins' reasoning. Nozick concluded that "if the actions or decisions to do the actions do not affect, help bring about, influence, etc., *which* state obtains, then whatever the conditional probabilities . . . one should perform the dominant action." He thus recommends taking both boxes.

Must It Be a Hoax?

Martin Gardner made the interesting claim that the type of prediction required is impossible: Any real-world Newcomb experiment must be a hoax, or the evidence of the predictor's accuracy must be invalid. Were he ever faced with a real Newcomb experiment, Gardner said, it would be "as if someone asked me to put 91 eggs in 13 boxes, so each box held seven eggs, and then added that an experiment had proved that 91 is prime. On that assumption, one or more eggs would be left over. I would be given a million dollars for each leftover egg, and 10 cents if there were none. Unable to believe that 91 is prime, I would proceed to put seven eggs in each box, take my 10 cents and not worry about having made a bad decision."

If the experiment as stated is inherently impossible, it changes everything. No prediction means no paradox, and you certainly should take both boxes. Still, the *practical* difficulties of performing the experiment should be irrelevant. Even whether there is or is not such a thing as ESP or an omniscient being is probably beside the point. The question is whether there is any possible way of effecting that type of prediction. It could be that there is something self-contradictory about prediction of another person's actions (especially where the person knows that his actions have been predicted).

No one can predict arbitrary human actions with the accuracy of Newcomb's paradox. This is rarely cited as a fundamental flaw in the situation, however. The idea that the human body, including the brain, is subject to the same physical laws as the rest of the universe is accepted as a commonplace in both the scientific and philosophic communities. If human actions are deterministic, then we must be open to the possibility of predicting them.

It seems to me that a Newcomb experiment could be carried out in practice. The method I propose is a frank cheat, but perhaps it does not fundamentally change the situation. Let the psychic be a fake who uses unknown trickery to accomplish the feat. The trickery need not (and must not) violate the rules. Possibly the psychic has discovered that, after mulling over the situation, 90 percent of the general public invariably takes box B only. In that case, he *always* predicts the subject will take B only, and he is right the claimed 90 percent of the time.

After discussing the paradox in a 1973 issue of *Scientific American*, Martin Gardner reported that the people writing to the magazine favored taking box B only by a margin of 2.5 to 1. If the correspondents were typical, then *anyone* could predict correctly more than 70 percent of the time by always saying the subject will take box B only. A 70 percent accuracy is well above the 50.05 percent threshold needed for the paradox with amounts of $1000 and $1 million. There is even enough slack for a cagey "psychic" to throw in an occasional prediction of both boxes to throw onlookers off the track.

The subjects must, of course, remain ignorant of this method of "prediction." In view of the success of many fake psychics (who likewise conceal their methods from their subjects), I think it possible that a charlatan could attain a track record of correct predictions and allow a Newcomb experiment.

Nonetheless, there is a larger and more interesting question of whether something as complex as human behavior can be predicted. Human beings are capable of defying predictions.

Two Types of Prediction

Science is good at predicting *some* things. Eclipses for the year 5000 A.D. can be predicted with certainty and relative ease. The morning weather report is often wrong by the afternoon. Why the disparity?

Evidently some phenomena are more predictable than others. This stems from the fact that there are two kinds of prediction. One variety uses modeling or simulation. You create a representation of the subject of the prediction that is as complex as the subject itself. The other, simpler kind of prediction uses "shortcuts" to accomplish the same thing.

What day of the week will it be 100 days from now? A calendar typifies the modeling approach. Each of the 100 future days is represented by a square of paper on a leaf of the calendar. Count forward 100 days, and read off the answer.

You could also recognize this shortcut: Divide 100 by 7, and take the remainder. It will be *that* many days of the week from the current day. One hundred divided by 7 leaves a remainder of 2. If today is Monday, two days from now will be Wednesday. One hundred days from now will also be Wednesday.

Whenever possible, we prefer the shortcut method. What if you wanted to know the day of the week 1,000,000 days from now? There may not be any calendar on earth that covers that day. You would have to make your own calendars for the next several thousand years. The shortcut method avoids that kind of busywork. Dividing 1,000,000 by seven and taking the remainder is scarcely more difficult than dividing 100.

Unfortunately, we are often forced to resort to a model. Some phenomena allow no shortcuts in their prediction. No method, no model that is any simpler than the phenomenon itself, will predict it.

Chaos

Blow up a toy balloon without tying it, then let go. The balloon's wild path around the room is unpredictable. Could you, by measuring the exact position and degree of inflation of the balloon at the moment of release, predict its path? Probably not. No matter how accurate your measurements, they wouldn't be accurate enough.

Determining the initial state of the balloon and room entails a lot more information than has been mentioned here. The pressure, temperature, and velocity of the air at each point in the room would have to be known, for the balloon interacts with the air it passes through. Eventually the balloon will bump against walls or furniture, so an exact knowledge of everything in the room would be necessary.

Even this knowledge would fall short. The balloon would still go this way and that, and end up in a different spot each time it is released. This failure of prediction is remarkable in a way. The balloon does not invoke unknown laws of physics. Its motion is a matter of air pressure, gravity, and inertia. If we can predict the orbit of Neptune millennia into the future, why do we fail with a toy balloon?

The answer is *chaos*. This is a relatively new term for phenomena that are unpredictable though deterministic. Science mostly deals in the predictable. Yet the unpredictable is all around us: a crack of lightning, the spurting of a bottle of champagne, the shuffling of a deck of cards, the meandering of rivers. There is reason to consider chaos the norm and predictable phenomena the freaks.

"Random" phenomena are governed by the same physical laws as everything else. What makes them unpredictable is this fact: In chaotic phenomena, the error in our measurement of their initial state grows exponentially with time. Jules Henri Poincaré anticipated chaos when he wrote in 1903:

A very small cause which escapes our notice determines a considerable effect that we cannot fail to see, and then we say that the effect is due to chance. If we knew exactly the laws of nature and the situation of the universe at the initial moment, we could predict exactly the situation of that same universe at a succeeding moment. But even if it were the case that the natural laws had no longer any secret for us, we could still only know the initial situation *approximately.* If that enabled us to predict the succeeding situation with *the same approximation,* that is all we require, and we should say that the phenomenon had been predicted, that it is governed by laws. But it is not always so; it may happen that small differences in the initial conditions produce very great ones in the final phenomena. A small error in the former will produce an enormous error in the latter. Prediction becomes impossible, and we have the fortuitous phenomenon.

Any measurement is a little bit off. If your driver's license says you are six feet one inch tall, it doesn't mean you are precisely that tall. The measurement was rounded off to the nearest inch; the measuring stick warped slightly since it was calibrated; you slouched a little during the measurement; your height has changed minutely since the measurement. You accept the fact that a measurement of human height is liable to be as much as 1 percent off, and leave it at that. We can live with such errors of measurement because they don't increase. In other contexts, a small error compounds until it is so huge we no longer have any knowledge of the measured quantity.

Chaos is the unstated rationale behind shuffling cards. After a hand of poker, the dealer collects the cards to shuffle. Inevitably, some people see which cards go where in the reassembled deck. One person notes the two of spades on the bottom; another sees that his hand, a straight, went on top. Each person has some knowledge, and some uncertainty, about the composition of the deck. Shuffling multiplies this uncertainty.

Suppose you had a straight flush, the 6-7-8-9-10 of hearts, which you arranged in that order, and saw that the dealer picked up the hand intact while reassembling the deck. If the dealer *didn't* shuffle before dealing, this would give you information about the other players' new hands. If in the next hand you were dealt the 8 of hearts, you could conclude that the player before you in the deal got the 7 of hearts, the player after you got the 9 of hearts, and so on.

On the average, a riffle shuffle inserts one card between cards that were originally adjacent. The sequence 6H-7H-8H-9H-10H becomes 6H-?-7H-?-8H-?-9H-?-10H and then 6H-?-?-?-7H-?-?-?-8H-?-?-?-9H-

?-?-?-10H. The space between originally adjacent cards doubles with each shuffle. By the second shuffle the first and last cards of the original straight flush are sixteen cards apart; chances are good that the cut for the third shuffle will split them into different packets. Then the cards will be scattered throughout the deck.

This understates the confusion, for of course no one shuffles with perfect interleaf. Sometimes two cards fall instead of one; sometimes several cards fall. The uncertainty about the shuffling process hikes the total uncertainty with each shuffle. Try this experiment: Put the ace of spades on top of a deck and riffle shuffle a few times. The ace of spades quickly migrates down through the deck. (It may remain on top a few shuffles depending on how the cards fall.) Were the deck infinitely large, the position of the ace of spades from the top would approximately double with each shuffle. So too would any small uncertainty about the card's position double with each shuffle. In a finite deck, once the card is shuffled below the midpoint of the deck, it goes in the bottom packet on the next shuffle and thereafter can be anywhere in the deck. To completely lose the card in a standard deck requires about six or seven shuffles.

Chaotic phenomena are said to be *irreducible*. They cannot be reduced to a model that is any simpler than themselves. A "model" can be many things: an equation, a working scale model, the set of neural circuits in your brain corresponding to your thoughts about the phenomenon. A stable orbit can be represented by a few equations or a planetarium. It is impossible, however, to build a shoe box-size scale model of a deflating balloon in a room so precise it allows prediction of where the full-sized balloon in the full-sized room will go. It is all the more impossible to model fully a river, a tornado, or a brain. The simplest representation of a chaotic phenomenon is the phenomenon itself. There is more to a cuckoo than a cuckoo clock can contain.

The irreducibility of the brain is suggested by this experiment: Think of an obscure past experience; think of a person you were with at the time, someone you have not thought of in a long time; count the number of letters in that person's first name; then, *if and only if* that number is odd, dog-ear the corner of this page. Could even your closest friend predict whether you will dog-ear the page? In a situation like this (and in many others), one minute part of your memory, a few neurons perhaps, can zoom to the forefront and determine your train of thought. No one could hope to predict what you would do in that situation unless they shared all your memories in cellular and even molecular detail. Nothing simpler than yourself can be expected to behave exactly as you would.

Chaos is distinct from quantum uncertainty. A world made of fully deterministic atoms would still have chaos. Together, chaos and quantum uncertainty make prediction all the more difficult. Even in ideal situations where there are no other sources of error, there is always quantum uncertainty. Chaotic phenomena magnify it over and over. Quantum uncertainty bubbles up to the everyday world and renders it unpredictable.

Free Will vs. Determinism

Philosophers make much of the conflict between free will and determinism. How can there be free will in a deterministic world? This question has occupied philosophers ever since the mechanistic philosophy gained sway. It is a big part of the puzzle of Newcomb's paradox.

There are at least three ways of dealing with the question. You may decide that there is no such thing as free will, and that's that. Free will is an illusion.

The trouble with that is, everyone feels like he has free will in most things. In plain everyday life, not having free will means you want to do something and some outside agency prevents it. You want to speak your mind about the Premier, but here in Transylvania they send you to the salt mines if you do that. You probably *wouldn't* feel your free will compromised if you learned that the states of quarks and gluons in your brain are strictly determined by physical law.

Alternatively, you can say that determinism is the illusion. The world, or at least the human mind, is not completely determined by the past. This option is unattractive to most contemporary thinkers. You have to turn your back on the science of the past five centuries to deny that events are constrained by natural law (quantum theory notwithstanding) and don't just happen any which way.

The "compatibilist" position says that there is no essential contradiction between free will and determinism. Determinism does not necessarily imply predictability (much less absence of free will). Our growing appreciation of the role of chaos in the universe lends a plausibility to this position.

Free will means doing as you please, even if what you please is predetermined by the states of neurons in your head. If your actions are predetermined but neither you nor anyone else can ever learn what is going to happen before it happens, the seeming conflict is avoided. You might well ask what difference that kind of determinism makes. The future is still unknowable. Do what you will, no one is ever looking over your shoulder and muttering with certainty, "Yep, he's going to take both boxes."

The only way that determinism can impinge on our sense of free will is when we learn of our predestination. Presumably God knows whether or not you will squeeze the toothpaste tube from the middle tomorrow morning. No problem—as long as God doesn't tell you. The unacceptable case is *knowing* that you are destined to make such and such a choice, and being "forced" by all those unfeeling atoms to do it. Only then is deterministic physical law the sort of coercive agency that prevents us from having free will.

Prediction and Infinite Regress

The problems of predicting irreducible phenomena are many. One thought experiment sometimes advanced in connection with Newcomb's paradox goes like this: A hermetically sealed chamber contains a super-computer with an exact knowledge of all the atoms in the room. All the laws of physics, chemistry, and biology are programmed into the computer, so it can predict anything that is going to happen in the room. (The room must remain sealed so that outside agencies do not affect any prediction.) A terrarium in the room contains a few frogs and plants. The computer predicts the births, deaths, matings, territorial struggles, and mental states of the frogs: All these predictions amount to charting the courses of a large but finite number of atoms in the terrarium. No light bulb can burn out, no coat of paint can blister without the computer predicting it.

In the room are also several people. Again, the computer knows every atom of their makeup. One person gets annoyed at the seeming violation of free will and asks the computer this question: "Will I be standing on my head at midnight tonight?" She then announces: "Whatever the computer predicts, I'm going to do the opposite. If it says I *am* going to be standing on my head at midnight, then I will do everything within my power to make sure that I don't stand on my head. If it says I won't stand on my head, I will." What would happen in this situation?

There are ways the computer could avoid being wrong. It could decline to answer, not answer until one minute after midnight, or answer in a language unknown to the room's inhabitants. It could predict the subject would not be standing on her head, and she could fall asleep early that night, forgetting the whole thing. But just because such scenarios avoid paradox does not mean they must occur.

Assuming a timely prediction is forthcoming, there is no reason why the person *couldn't* make good on her promise. Call free will an illusion if you like; any of us can resolve to do a headstand (or not). The computer's prediction is not going to make anyone any less free to do so.

In fact, the computer cannot make a valid prediction. To see why, ask how the computer makes its prediction. Does it use a shortcut: a rule, a gimmick, a mathematical formula? It is beyond belief that any *simple* rule tells whether a specific person will be standing on her head at a given moment! It is one thing to predict days of the week, seasons, or comet returns. There is regularity in these phenomena. There is no regularity to standing on one's head. Even if there was (if the subject was in the habit of standing on her head every second Tuesday at midnight), her promise to be contrary invalidates it.

Evidently, the computer predicts by modeling the situation in the room. It was stated that the computer predicted the very courses of the atoms to foresee the actions of the frogs. Here we come to the crux of the paradox. Since the subject certainly will be influenced by the computer's prediction, the computer must predict its own prediction as well as the subject's reaction to it. The computer's model must represent the computer itself *in full detail.*

This paradoxical requirement recalls the map Borges and Adolfo Bioy Casares describe in *Extraordinary Tales:*

> In that empire, the art of cartography achieved such perfection that the map of one single province occupied the whole of a city, and the map of the empire, the whole of a province. In time, those disproportionate maps failed to satisfy and the schools of cartography sketched a map of the empire which was the size of the empire and coincided at every point with it. Less addicted to the study of cartography, the following generations comprehended that this dilated map was useless and, not without impiety, delivered it to the inclemencies of the sun and of the winters. In the western deserts there remain piecemeal ruins of the map, inhabited by animals and beggars. In the entire rest of the country there is no vestige left of the geographical disciplines.

The computer wants to set aside a certain portion of its available memory to simulate its own actions. Unfortunately, no part of the computer's workings any smaller than the whole computer can do this. The most efficient way the computer can model itself is to *be* itself. That, like Borges and Casares's map, leaves no room for anything else.

Even if the computer is highly redundant, you run into trouble. Some computers, such as those used for spacecraft navigation and life support, have two or more separate subsystems doing the same thing. This greatly reduces the chance of an error. Potentially, it also allows each of the redundant subsystems to "predict" what the computer as a whole will do.

Compare this to a Borges-Casares map at a scale of 1:2. The map is half as wide as the country it represents. A 1:2 map of the United States stretches from San Francisco to Kansas City, blanketing the mountain states. A map that big is itself a significant man-made feature worthy of inclusion on all maps of the country. That means the 1:2 map must show itself. And the map on the map must contain a map of itself, and so on, ad infinitum.

For the same reason, a redundant computer modeling itself would contain a model of the computer, a model of the model, a model of the model of the model . . . Nothing prevents you from imagining this. But no real computer made of atoms can attain an infinite regress. The models and models in models must have some physical reality as states of memory chips, and memory chips cannot be infinitely small. The prediction is therefore impossible.

Now back to Newcomb's situation. You can make a good case that the experiment as usually stated is impossible—that is, the prediction is impossible. The reason is essentially the same as above. Endless regress rules out a 100 percent accurate prediction.

Yes, but wouldn't we be satisfied with a 90 percent prediction? Given the contrary nature of the human mind, small uncertainties about a subject's or predictor's mental state may grow exponentially and produce total uncertainty. Predicting the system of predictor and predicted is as impossible as predicting any chaotic phenomenon. Thus there is no casual distinction between 100 percent and 90 percent accuracy. It is like being 90 percent sure of the arrangement of a deck of cards before it was shuffled. Unless the predictor can get a complete lock on the state of the room (it can't), it may not be able to predict with *any* accuracy at all.

If the paradoxes I have gathered on these pages have one recurring theme, it is the folly of denying ignorance. Just because something is so doesn't mean that we can know it. There is necessary ignorance, and it is more significant than mere solipsism would have us believe.

The assumption that anything true is knowable is the grandfather of paradoxes. In its purest form this is the foundation of Buridan sentences and infinity machines. The riddles of Hempel and Goodman trade on the fallacy that the import of any observation is knowable a priori. The victim of the unexpected hanging errs in thinking he can deduce something he cannot; Newcomb's experiment founders on the impossibility of a predictor knowing his own mind.

The physicist Ludwig Boltzmann conjectured that our astonishment at the order of the world is misplaced: The known universe may be one small random fluctuation in an infinite universe that contains all possible arrangements of atoms. One may be forgiven for wondering if our knowledge is similarly engulfed in a larger whole. Perhaps the real mystery is that everything imaginable is true—somehow, somewhere—and our minds are preoccupied with an infinitesimal part of the whole of existence, whose paths we have worn in our first explorations.

Newcomb's Paradox 3000 A.D.

There is no more satisfying resolution to a decision paradox than to show that the situation itself can never arise. Unfortunately, I am forced to conclude that a slightly modified version of the experiment *is* conceivable. To do so, I resort to either of two science-fictional devices.

The Newcomb experiment is being held in the year 3000 A.D., and the predictor has at his disposal two gadgets, a time machine and a matter scanner. In the first case, the predictor hops in the time machine and sets the dial for just after the subject makes his choice. Arriving in the near future, he gets out and learns the decision. Then he gets back in the time machine and returns to the day before the experiment. He makes his prediction on the strength of certain knowledge of the future.

This would give the subject planning to take both boxes pause to consider. You're the subject, and you notice a video camera in the corner of the room. Just before you decide, the predictor walks in and hands you a videotape. *It's a tape of you opening the box(es) he brought back from the future.* Not only has the predictor an evidently correct prediction, he has a motion picture of it.

Time travel is such an iffy idea that it may be unwise to set much store by it. The other futuristic method of prediction, one you might be more comfortable with, is a matter scanner. The scanner duplicates matter exactly. You set up the machine, scan a $1000 bill, and it creates a new bill that is identical in every possible way, down to the quantum states of its atoms. Scanning a person with the machine creates an exact duplicate. Armed with this technology, it is also possible to predict with certainty the result of a Newcomb experiment.

There are logistic subtleties, though. It would not do simply to create a twin of the subject and do a trial run of the experiment with the twin. The minutiae of the experiment would not be the same. It would be a different time of day; the twin might be in a different mood; the experimental proctor might explain something slightly differently. These trifles might not make a difference, but you never know. The subject *could* be contrary. Knowing about the doppelgänger, he might decide to do just the opposite of his "first impulse" to prove his free will. We want to assure ourselves that the prediction can be arbitrarily certain.

To ensure the validity of the prediction, two drastic steps are necessary. You would have to create an exact duplicate of the subject, the boxes, the table, the room, the guards, everything and everyone involved in the experiment. The duplicated region would have to be so large that no outside influence could reach the subject before making the choice. You'd want a hermetically sealed, artificially lighted room. Otherwise, even the angle of the sunlight streaming through the windows might make a difference—and you can't duplicate the sun.

The other problem is timing. Now you have two subjects in two rooms. You want to fast-forward the duplicate experiment to its conclusion so that you know what the real subject will do before he does it. Otherwise, all this effort goes to waste; there is no prediction.

There are two conceivable solutions. One, you could pack the original subject and surroundings into a gigantic rocket and shoot them away from the earth at near the speed of light. The rocket's computers steer a course several light-years out into space, turn around, and return at near the speed of light. The rocket acceleration creates 1 g of artificial gravity, so the subject in the sealed room remains ignorant of his journey. By the time the subject returns in the rocket, you know what the duplicate did; thanks to a twin paradox, the real subject has not yet made his choice.

A more practical method is to run the experiment with the duplicate as the subject. The scanning and duplicating process must take some time. So scan the original subject, watch what choice he makes, *then* create the duplicate. You predict what the duplicate will do.

All right then. It is the year 3000 A.D., and you are the subject in a Newcomb experiment. Just before you decide, you are told—gasp! —that you may be only a duplicate of the real you, created just five minutes ago. How Russellian! You have no reason to doubt this: Matter scanners are as common in 3000 A.D. as microwave ovens are today. You accept that the experiment's predictor attains 100 percent accuracy by watching an exact double in an identical room.

You might ask how you can possibly know if you are the duplicate or the original. You can't. The situation *is* exactly as in Russell's thought experiment about the world being created five minutes ago. Duplicate and original have identical memories, including the memory of walking to the room a few minutes ago so they could scan it and create the duplicate. Both original and duplicate are asked to make the choice of boxes.

The experiment's sponsors even had to tell the *real* you that he might be a duplicate too. The experiment is designed around you, the duplicate, who makes his choice after the original has been observed. Consequently the sponsors had to tell the original he might be a duplicate in order to tell you that you are. Only after choosing, when you walk out of the room, do you find out whether you are the original or the duplicate.

This nips contrary stratagems in the bud. You can't outsmart the predictor by doing one thing in the "trial run" and another in the real experiment. You have no way of knowing which is which. This is the case even when you have full knowledge of the prediction method.

Another technical point is what to put in the boxes. Whatever is inside the original room's boxes will necessarily be in the cloned room's boxes. Of course, they don't know what to put in the boxes until after the original room's experiment is run. The solution is to leave the boxes empty or eliminate them altogether. Instead, you state what you would do, and collect your money when you leave the room, provided you turn out to be the duplicate (the subject of the "real" experiment).

The matter scanner does not change the paradox one whit; it only demonstrates a way the prediction could avoid infinite regress. You can't dismiss Newcomb's paradox as depending on infinities, an omniscient deity, ESP, or any other imponderable. Agreed, quantum uncertainty would probably rule out the matter scanner. It seems unsatisfying, however, for mere physics to stand in the way of a paradox of logic.

If a matter scanner is possible, we would have the paradox in its acutest form. There would be two identical persons in identical rooms; they would, after puzzling over the situation, make their choice halfheartedly or with confidence; and the predictor, watching the first person, would know what the time-delayed doppelgänger would do as surely as he could know the outcome of a rerun football game on television. The subjects taking both boxes would always get $1000; the subjects taking box B would always get $1 million. There is the situation, and it is as inscrutable as ever.

Bibliography

THIS BOOK is only a sampler of the many provocative paradoxes and thought experiments being discussed in the scientific and philosophical literature. Those interested in further reading would do well to start with recent issues of *Analysis, The British Journal for the Philosophy of Science, Mind, Philosophical Studies,* and *Philosophy of Science.*

Bacon, Roger(?). The Voynich "Roger Bacon" Cipher Manuscript. Central Europe, probably sixteenth century. At Yale University's Beinecke Rare Book and Manuscript Library, New Haven.

Barber, Theodore Xenophon, and Albert Forgione, John F. Chaves, David S. Calverley, John D. McPeake, and Barbara Bowen. "Five Attempts to Replicate the Experimenter Bias Effect," *Journal of Consulting and Clinical Psychology,* 33:1–6 (1969).

Bennett, William Ralph, Jr. *Scientific and Engineering Problem-solving with the Computer.* Englewood Cliffs, N.J.: Prentice-Hall, 1976.

Borges, Jorge Luis. *Labyrinths: Selected Stories and Other Writings,* ed. by Donald A. Yates and James E. Irby. New York: New Directions, 1964.

———. *Other Inquisitions: 1937–1952.* Austin: University of Texas Press, 1964.

———. *A Personal Anthology.* New York: Grove Press, 1967.

———. *The Book of Sand.* New York: E. P. Dutton, 1977.

——— and Adolfo Bioy Casares, *Extraordinary Tales.* London: Condor Books, 1973.

Burge, Tyler. "Buridan and Epistemic Paradox," *Philosophical Studies,* 39:21–35 (1978).

Carroll, Lewis. *Symbolic Logic,* ed. by William Warren Bartley III. New York: Clarkson N. Potter, 1977.

266   BIBLIOGRAPHY

Coate, Randoll, Adrian Fisher, and Graham Burgess. *A Celebration of Mazes.* St. Albans, Eng.: Minotaur Designs, 1986.

Cole, David. "Thought and Thought Experiments," *Philosophical Studies,* 45:431–444 (1984).

Cook, Stephen. "The Complexity of Theorem Proving Procedures," *Proceedings of the 3rd Annual ACM Symposium on the Theory of Computing.* New York: Association of Computing Machinery, 1971.

Einstein, Albert, and Leopold Infeld. *The Evolution of Physics.* New York: Simon & Schuster, 1938.

Gardner, Martin. *The Scientific American Book of Mathematical Puzzles and Diversions.* New York: Simon & Schuster, 1959.

———. *The Unexpected Hanging and Other Mathematical Diversions.* New York: Simon & Schuster, 1969.

———. *Knotted Doughnuts and Other Mathematical Entertainments.* New York: W. H. Freeman, 1986.

Garey, Michael R., and David S. Johnson. *Computers and Intractability: A Guide to the Theory of NP-Completeness.* New York: W. H. Freeman, 1979.

Gettier, Edmund. "Is Justified True Belief Knowledge?" *Analysis,* 23:121–123 (1963).

Goodman, Nelson. *Fact, Fiction, and Forecast.* Indianapolis: Bobbs-Merrill, 1965.

Grünbaum, Adolf. "Are 'Infinity Machines' Paradoxical?" *Science,* 159:396–406 (Jan. 26, 1968).

Hazelhurst, F. Hamilton. *Gardens of Illusion: The Genius of André le Nostre.* Nashville: Vanderbilt University Press, 1980.

Heller, Joseph. *Catch-22.* New York: Simon & Schuster, 1961.

Hesse, Mary. "Ramifications of 'Grue,'" *British Journal for the Philosophy of Science,* 20:13–25 (1969).

Hofstadter, Douglas R. *Gödel, Escher, Bach: An Eternal Golden Braid.* New York: Basic Books, 1979.

——— and Daniel C. Dennett. *The Mind's I.* New York: Basic Books, 1981.

Hume, David. *A Treatise of Human Nature.* New York: Penguin, 1986.

Jevons, Stanley. *The Theory of Political Economy.* London, 1911.

Karp, Richard. "Reducibility among Combinatorial Problems," in R. E. Miller and J. W. Thatcher, eds., *Complexity of Computer Computations.* New York: Plenum Press, 1972.

Ladner, R. E. "On the Structure of Polynomial Time Reducibility," *Journal of the Association of Computing Machinery,* 22:155–171 (1975).

Leibniz, Gottfried. *Monadology,* trans. by Paul and Anne Schrecker. Indianapolis: Bobbs-Merrill, 1965.

Levitov, Leo. *Solution of the Voynich Manuscript: A Liturgical Manual For The Endura Rite Of The Cathari Heresy, The Cult of Isis.* Laguna Hills, Calif.: Aegean Park Press, 1987.

Olds, James. "Pleasure Centers in the Brain," *Scientific American,* Oct. 1956, pp. 105–116.

Penfield, Wilder. "The Cerebral Cortex in Man," *Archives of Neurology and Psychiatry,* 40:3 (Sept. 1938).

Plato. *The Dialogues of Plato,* trans. by B. Jowett. New York: Random House, 1937.

Putnam, Hilary. *Mind, Language and Reality.* New York: Cambridge University Press, 1975.

———. *Reason, Truth and History.* New York: Cambridge University Press, 1981.

Rado, Tibor. "On Non-Computable Functions," *The Bell System Technical Journal,* May 1962.

Rescher, Nicholas, ed. *Essays in Honor of Carl G. Hempel.* Dordrecht, Holland, 1969.

Rosenthal, Robert. *Experimenter Effects in Behavioral Research.* New York: Appleton-Century-Crofts, 1966.

Rucker, Rudy. *Infinity and the Mind: The Science and Philosophy of the Infinite.* Cambridge, Mass.: Birkhauser, 1982.

Russell, Bertrand. *The Principles of Mathematics.* London: Allen and Unwin, 1937.

———. *Human Knowledge: Its Scope and Limits.* New York: Simon & Schuster, 1948.

Salmon, Wesley, ed. *Zeno's Paradoxes.* New York: Irvington, 1970.

Searle, John. "Minds, Brains, and Programs," *Behavioral and Brain Sciences* 3:442–444 (1980).

Smullyan, Raymond. *What Is the Name of This Book? The Riddle of Dracula and Other Logical Puzzles.* Englewood Cliffs, N.J.: Prentice-Hall, 1978.

———. *This Book Needs No Title: A Budget of Living Paradoxes.* Englewood Cliffs, N.J.: Prentice-Hall, 1980.

Turing, Alan M. "Computing Machinery and Intelligence," *Mind,* 59, no. 236 (1950).

Vonnegut, Kurt, Jr. *Cat's Cradle.* New York: Delacorte Press, 1963.

Walker, Jearl. "Methods for Going Through a Maze Without Becoming Lost or Confused," *Scientific American,* Dec. 1986.

Watkins, Ben. *Complete Choctaw Definer.* Van Buren, Ark.: J. W. Baldwin, 1892.

Whitrow, G. J. "On the Impossibility of an Infinite Past," *British Journal for the Philosophy of Science,* 29:39–45 (1978).

Index